Tributes
Volume 35

Language, Evolution and Mind
Essays in Honour of Anne Reboul

Volume 25
The Facts Matter. Essays on Logic and Cognition in Honour of Rineke Verbrugge
Sujata Ghosh and Jakub Szymanik, eds.

Volume 26
Learning and Inferring. Festschrift for Alejandro C. Frery on the Occasion of his 55th Birrthday
Bruno Lopes and Talita Perciano, eds.

Volume 27
Why is this a Proof? Festschrift for Luiz Carlos Pereira
Edward Hermann Haeusler, Wagner de Campos Sanz and Bruno Lopes, eds.

Volume 28
Conceptual Clarifications. Tributes to Patrick Suppes (1922-2014)
Jean-Yves Béziau, Décio Krause and Jonas R. Becker Arenhart, eds.

Volume 29
Computational Models of Rationality. Essays Dedicated to Gabriele Kern-Isberner on the Occasion of her 60th Birthday
Christoph Beierle, Gerhard Brewka and Matthias Thimm, eds.

Volume 30
Liber Amicorum Alberti. A Tribute to Albert Visser
Jan van Eijck, Rosalie Iemhoff and Joost J. Joosten, eds.

Volume 31
"Shut up," he explained. Essays in Honour of Peter K. Schotch
Gillman Payette, ed.

Volume 32
From Semantics to Dialectometry. Festschrift in Honour of John Nerbonne.
Martijn Wieling, Martin Kroon, Gertjan van Noord, and Gosse Bouma eds.

Volume 33
Logic and Computation. Essays in Honour of Amílcar Sernadas
Carlos Caleiro, Fransciso Dionísio, Paula Gouveia, Paulo Mateus and João Rasga, eds.

Volume 34
Models: Concepts, Theory, Logic, Reasoning, and Semantics. Essays Dedicated to Klaus-Dieter Schewe on the Occasion of his 60th Birthday
Atif Mashkoor, Qing Wang and Bernhrd Thalheim, eds.

Volume 35
Language, Evolution and Mind. Essays in Honour of Anne Reboul
Pierre Saint-Germier, ed.

Tributes Series Editor
Dov Gabbay   dov.gabbay@kcl.ac.uk

Language, Evolution and Mind
Essays in Honour of Anne Reboul

edited by
Pierre Saint-Germier

© Individual authors and College Publications 2018. All rights reserved.

ISBN 978-1-84890-282-4

College Publications
Scientific Director: Dov Gabbay
Managing Director: Jane Spurr

http://www.collegepublications.co.uk

Cover design by Laraine Welch

Printed by Lightning Source, Milton Keynes, UK

All rights reserved. No part of this publication may be reproduced, stored in a retrieval system or transmitted in any form, or by any means, electronic, mechanical, photocopying, recording or otherwise without prior permission, in writing, from the publisher.

Language, Evolution and Mind
Essays in Honour of Anne Reboul

edited by

Pierre Saint-Germier

© Individual authors and College Publications 2018. All rights reserved.

ISBN 978-1-84890-282-4

College Publications
Scientific Director: Dov Gabbay
Managing Director: Jane Spurr

http://www.collegepublications.co.uk

Cover design by Laraine Welch

Printed by Lightning Source, Milton Keynes, UK

All rights reserved. No part of this publication may be reproduced, stored in a retrieval system or transmitted in any form, or by any means, electronic, mechanical, photocopying, recording or otherwise without prior permission, in writing, from the publisher.

Contents

1	**Introduction** *Pierre Saint-Germier*	1
I	**Linguistics and Philosophy of Language**	**11**
2	**The Semantics-Pragmatics Interface:** **How It Works, Why We Need It and Where It Is** *Jacques Moeschler*	13
3	**The Semantics of Functional Prepositions in Serbian** *Tijana Ašić*	39
4	***De Se* Readings as a Window on the Nature of Control** *Denis Delfitto & Gaetano Fiorin*	51
5	**A Pragmatic Promenade in the French Landscape of Colours** *Louis de Saussure*	87
6	**How Logical is Natural Language Conjunction? An Experimental Investigation of the French Conjunction *et*** *Joanna Blochowiak & Thomas Castelain*	97
7	**Contexts, Biases, and Reflexivity** *Eros Corazza*	127

8 Metaphor, Relevance and the Interpretation of Fiction   149
   Deirdre Wilson

II Language evolution   169

9 Language Evolution: Insisting on Making It a Mystery or Turning
   It into a Problem?   171
   Pedro Tiago Martins & Cedric Boeckx

10 On Language not Being at Root a Communication System.
   Some Morphosyntactic Considerations   181
   Frederick J. Newmeyer

III Cognitive Science and Philosophy of Mind   201

11 Stop Moving and You Will Understand. Selective Weakening of
   the N400 to Hand-Action Words during Hand Movements   203
   *Tatjana A. Nazir, Raphaël Fargier, Evgueni Douissembekov
   & Yves Paulignan*

12 Should Embodied Cognitive Science Go Radical?
   A Hint from Music   217
   Pierre Saint-Germier

13 Preschoolers' Social Preferences in a Dominance Context   247
   *Rawan Charafeddine, Chloé Billamboz, Ira Noveck
   & Jean-Baptiste Van der Henst*

14 How to Be a Direct Realist   263
   Alfredo Paternoster

15 Intentional Imagination and Delusion   279
   Philip Gerrans & Kevin Mulligan

Contributors

Tijana Ašić
University of Kragujevac and University of Belgrade

Chloé Billamboz
Laboratoire L2C2, Institut des Sciences Cognitives Marc Jeannerod, Lyon

Joanna Blochowiak
University of Geneva

Cedric Boeckx
Universitat de Barcelona, ICREA, UBICS

Thomas Castelain
Universidad de Costa Rica, Instituto de Investigaciones Psicológicas

Rawan Charafeddine
Laboratoire L2C2, Institut des Sciences Cognitives Marc Jeannerod, Lyon

Eros Corazza
Institute for Logic, Cognition, Language and Information, Universidad del Pais Vasco, Ikerbasque, and Carleton University

Denis Delfitto
Università degli studi di Verona

Evgueni Douissembekov
Laboratoire L2C2, Institut des Sciences Cognitives Marc Jeannerod, Lyon

Raphaël Fargier
Laboratoire L2C2, Institut des Sciences Cognitives Marc Jeannerod, Lyon

Gaetano Fiorin
Utrecht University

Philip Gerrans
University of Adelaide

PEDRO TIAGO MARTINS
Universitat de Barcelona, UBICS

JACQUES MOESCHLER
University of Geneva

KEVIN MULLIGAN
Università della Svizzera italiana, Lugano

TATJANA A. NAZIR
Laboratoire L2C2, Institut des Sciences Cognitives Marc Jeannerod, Lyon

FREDERICK J. NEWMEYER
University of Washington, University of British Columbia and Simon Fraser University

IRA NOVECK
Laboratoire L2C2, Institut des Sciences Cognitives Marc Jeannerod, Lyon

ALFREDO PATERNOSTER
Università degli studi di Bergamo

YVES PAULIGNAN
Laboratoire L2C2, Institut des Sciences Cognitives Marc Jeannerod, Lyon

PIERRE SAINT-GERMIER
Centre for Science Studies, Aarhus University

LOUIS DE SAUSSURE
Université de Neuchâtel

JEAN-BAPTISTE VAN DER HENST
Laboratoire L2C2, Institut des Sciences Cognitives Marc Jeannerod, Lyon

DEIRDRE WILSON
University College London and Centre for the Study of Mind in Nature, Oslo

Chapter 1

Introduction

Pierre Saint-Germier

The present volume originated in a collection of papers offered to Anne Reboul on the occasion of her 60th birthday and published online (Dupuy et al. 2016) to celebrate her academic career. It contains essays written by scholars who have interacted with her over the years and share a deep respect for her work as a linguist, cognitive scientist and philosopher. The tripartite structure of the volume roughly corresponds to three of the main subject-matters investigated by Reboul: language, evolution and mind. In this liminary chapter, I will present the organization of each part and briefly introduce the contents of the papers, to give the reader an overview of things to come.

The first part gathers chapters belonging to linguistics, broadly construed, and the philosophy of language. A majority of the contributions deal with issues of pragmatics, in line with Reboul's initial area of specialization (Moeschler & Reboul 1994; Reboul & Moeschler 1998a, 1998b). Syntax and Semantics are also represented, as well as Experimental Pragmatics, a field to which Reboul and her colleagues of the Institut des Sciences Cognitives Marc Jeannerod de Lyon were early contributors (e.g. Noveck & Reboul 2008).

Jacques Moeschler opens up with a chapter about "The Semantics-Pragmatics Interface: How It Works, Why We Need It and Where It Is". Its title gives a clear idea of the three questions that animate his contribution. Regarding the first one, he rejects both the view that pragmatic meaning is always dependent on truth-conditional meaning and the view that pragmatic meaning systematically intrudes into truth-conditional meaning: the SPI is a porous border in both directions. Second, he illustrates the function of the Semantics-Pragmatics Interface, i.e. allowing quick and efficient information transfer between linguistic and non-linguistic sources, with the pragmatic inferences triggered by quantifiers, analyzed as explicatures rather than scalar implicatures. Third, he provides a multi-layered explanation of the difference between the semantic and pragmatic meanings of the French causal connectives *parce que*, *donc* and *et*, and offers on this basis his own preferred way to locate the border between semantics and pragmatics.

In "The Semantics of Functional Prepositions in Serbian", Tijana Ašić investigates systematically the semantics of the Serbian functional prepositions *pred* and *za*. These prepositions denote spatial relations but also convey that the subject enters an active relationship with the prepositional argument, related to its function. For example, *pred* communicates that the subject is in front of the prepositional argument, e.g., a television set, but then also suggests that the subject is *watching* the television, rather than just standing in front of it. The question Ašić addresses is whether this functional aspect should be part of the basic semantic meaning of such prepositions. She gives a negative answer on the basis of cancellability arguments. The semantics she proposes for the functional prepositions *pred* and *za* only requires mereotopological predicates as primitives. The functional meanings only arise as implicatures, offering a new illustration of the powers of Grice's razor.

In their joint contribution "*De Se* Readings as a Window on the Nature of Control", Denis Delfitto and Gaetano Fiorin show how the phenomenon of Immunity to Error through Misidentification (IEM), originally highlighted by philosophers, sheds light on some poorly understood syntactical properties of control predicates. Starting from a paradox faced by an otherwise promising approach to the analysis of restructuring phenomena (Cinque 2004), they propose a solution based on the idea that the verbs featuring IEM trigger an overwriting of the lower theta-role by the higher theta-role. This proposal is independently motivated by reference the nature of the mental states denoted by those verbs and provides an enlightening explanation of some otherwise puzzling features of Partial Control, showing admirably how philosophical insights can provide valuable resources for solving technical difficulties in the theory of syntax.

Louis de Saussure's chapter invites the reader to a "Pragmatic Promenade in the French Landscape of Colours", an area also frequented by Reboul (2015a). Along the way, Saussure makes a number of striking observations regarding the morphological properties of French color adjectives and verbs. He suggests a pragmatic explanation for the intuitive oddity of suffixes that should be compatible with basic terms (as in *beigeâtre*), and the felicity, in some contexts, of combinations that should not make sense (such as *beigeâtre*). Even though the pragmatic story about colour terms is just one chapter in a much longer chronicle involving evolutionary, linguistic and cultural evolution, it is a story worth reading for itself—and full of surprises.

Joanna Blochowiak and Thomas Castelain's chapter "How Logical is Natural Language Conjunction? An Experimental Investigation of the French Conjunction *et*" represents in this volume the approach of experimental pragmatics. Blochowiak and Castelain evaluate some competing views about the processing of natural language conjunction, derived from pragmatics (Levinson 2000, Relevance Theory) and syntactic hypotheses (Bjorkman 2010). Blochowiak and Castelain designed a self-paced reading experiment to evaluate the relative processing cost of symmetric (i.e. logical) and asymmetric (i.e. temporal, causal) interpretations of the conjunction *et* in French. The results obtained indicate that symmetrical propositions are processed faster than asymmetric ones and that among the asymmetric ones, the temporal interpretation is processed faster than the causal ones. Blochowiak and Castelain raise interesting methodological issues about the proper interpretation of symmetrical readings of *et*, commonly assumed to indicate a logical relation, whereas the feature of symmetry is also shared by some non-logical readings (e.g. simultaneity and temporal overlapping). On a careful interpretation of the data, only Levinson's prediction that temporal interpretations are processed faster than causal ones gets a clear confirmation. Blochowiak and Castelain also suggest an alternative analysis in terms of explicatures, consistent with their data, which opens an interesting path for further empirical investigations.

In "Contexts, Biases, and Reflexivity", Eros Corazza discusses the theory of biases proposed by Stefano Predelli in his (2013) book. Predelli's strategy, in a nutshell, is to integrate non truth-conditional meaning, for example the derogatory meaning of slurs, to a truth-conditional semantic framework by conventionally associating to linguistic expressions a *bias*, i.e. a constraint on what counts as a proper context of use for these expressions, in addition to their Kaplanian character. For example, the slur "frog" comes with a bias that restricts its appropriate use to contexts where French people bear, or are thought by the speaker to bear, some negative traits typical of French people. Thus any speaker uttering:

(1)   Pierre is not French, he is a frog!

communicates that she takes French people to have the (allegedly) typical negative traits and thus expresses a kind of anti-French prejudice. But do we need to add this new component to the conventional meaning of words? Focusing on the case of slurs, Corazza examines the plausibility of alternative explanations. While he rejects an explanation of biases in terms of conventional implicatures, he defends an alternative account of the same phenomena in terms of generalized conversational implicatures integrated to a pluri-propositionalist approach of communication, drawing on the recent work of Korta and Perry (2011). The non truth-conditional meaning of slurs, and presumably of other puzzling expressions, can thus be integrated to a theory of communication without having to add biases.

In her contribution on "Metaphor, Relevance and the Interpretation of Fiction" Deirdre Wilson addresses some questions raised by Reboul (1986,1987) about the ability for Relevance Theory to deal with fiction—another favourite theme of Reboul's (Reboul 1992). According to Relevance Theory, the allocation of attention is seen as guided by expectations of relevance, that is by the expectation of a valuable cognitive effect, in exchange for an affordable processing cost. But it is not clear how relevance can be achieved without truth. Thus fictions raise a *prima facie* difficulty for Relevance Theory. Reboul's proposal was to consider the relevance of fictions as analogous to the relevance of metaphors—which raises essentially the same difficulty, at least if metaphors are interpreted in such a way that they do not generate truths, as they were initially in Sperber and Wilson (1986/1995). But the Relevance-Theoretic treatment of metaphor has evolved over the years. Wilson defends the more recent *ad hoc* concept account of metaphor, against some criticism of Reboul (2011), and draws some implications for the parallel with fiction: if the *ad hoc* concept view is adequate, then an analogy with metaphor might not be the best strategy to explain the relevance of fictions.

The second part of the volume deals specifically with the evolution of language, a field to which Reboul has recently contributed with her book *Cognition and Communication in the Evolution of Language* (Reboul 2017). In their chapter "Language Evolution: Insisting on Making It a Mystery or Turning It into a Problem?" Pedro Tiago Martins and Cedric Boeckx engage with an influential paper by Hauser et al. (2014) where the authors, after reviewing the results offered by comparative animal behaviour, paleontology and archaeology, molecular biology and evolutionary modelling, lament on the lack of progress made in the last decades of research on the evolution of language Martins and Boeckx forcefully disagree. They provide a critical examination of the arguments offered by Hauser et al. and draw a different, much less pessimistic conclusion.

Frederick Newmeyer's contribution "On Language not Being at Root a Communication System. Some Morphosyntactic Considerations" follows up a paper of Reboul (2015b), where she argues against the view that the driving force in the origin and evolution of language was the enhancement of communicative abilities. That view is consonant with some recent hypotheses (from Jackendoff, Progovac and Chomsky, respectively) according to which important features of morphosyntax were selected for incrementally on the basis of their enhancement of communication. Newmeyer raises a number of difficulties for these hypotheses regarding the evolution of morphosyntax. First, he argues that it is unlikely that the incremental steps towards morphosyntactical complexivity are attributable to biological evolution given that some of the evolved features are missing in otherwise equally expressive languages. Second, evidence of decremental evolution in some languages is hard to reconcile with the incremental model. Third, the apparition of some morphosyntactical features can be explained by alternative, non-biological, processes of language change. Some of Reboul's ideas regarding the evolution of language thus find some additional support from a morphosyntactic point of view.

The third part of the book shifts the focus from language to the mind, as investigated by cognitive scientists and philosophers of mind. The view that cognition is embodied has attracted a lot of attention in various areas of cognitive science. In cognitive neuroscience, it has been shown for example that processing sentences describing motor actions activates areas in the brain responsible for programming the execution of motor actions. In their chapter "Stop Moving and You Will Understand. Selective Weakening of the N400 to Hand-Actions Words during Hand Movements", Tatjana Nazir, Raphaël Fargier, Evgueni Douissembekov and Yves Paulignan investigate, using EEG and behavioural measures, the natural consequence that the ability to grasp the meaning of these sentences should decline when motor areas are occupied with planning and executing movements, while sentences that do not describe such actions should not be affected by the same movements. In one condition, they presented subjects with sentences ending either with a concrete noun or an action verb, this last word being either congruent or not congruent with the whole sentential context, and asked them whether they found the sentence congruent. In a second condition, they asked subjects to perform the same task while continuously grasping cylindrical object and place it in one of two holes. For the sentences ending with a verb, they observed that the semantic judgment errors are higher in the second condition than in the first one, which is consistent with the hypothesis under scrutiny. In the second condition, the "N400 effect", an electrophysiological marker for the processing efforts of semantically incongruous sentences, was smaller for sentences ending with verbs, but for not sentences ending with nouns. This indicates, consistently with the hypothesis, that the final verbs tend not to be processed because of the competing activation of the motor system. Nazir, Fargier, Douissembekov and Paulignan thus provide valuable new pieces of evidence in favour of the view that language processing and motor action share common brain resources.

Pierre Saint-Germier, in his chapter "Should Embodied Cognitive Science Go Radical? A Hint from Music" engages with the Radical Embodied Cognition advocated by Anthony Chemero. While "moderate" proponents of Embodied Cognition are happy to employ computational and representational explanations when dealing with the way the body of cognitive agents essentially contribute to their cognitive achievements, Chemero promotes the use of new tools, including Dynamical Systems Theory, and claims in addition that explanations of this new sort do not require to posit representations. Saint-Germier elaborates on a diagnosis offered by Reboul (2011) that this kind of dispute may turn out to be verbal and raises an additional difficulty for REC, drawing on some recent dynamical explanations of rhythm perception where both the *explanandum* and the *explanans* involve representational terms. This suggests that adopting dynamical explanations is not incompatible with positing representations and that some dynamical explanations do require representations to be explanatorily successful.

The contribution of Rawan Charafeddine, Chloé Billamboz, Ira Noveck & Jean-Baptiste Van der Henst "Preschoolers' Preferences in a Dominance Context", shifts from embodied to social cognition. Finding our way around in the social world requires distinctive cognitive capabilities such as the attributions beliefs, desires and feelings to others. Our social behaviour is also marked by preferences: we prefer to engage into joint actions with some individuals rather than others. Charafeddine, Billamboz, Noveck and Van der Henst investigate the way children understand relations of dominance among peers and how it affects their social preferences. The first experiment they present shows no preference of 4- and 5-years-old children for a dominant over or subordinate character, where the dominant character imposes their will on a subordinate in a situation of conflict. The second one compared the preference of 3-, 4- and 5-year-olds in a decisional power scenario similar to the one used in the first experiment and in a physical supremacy scenario, where the subjects were asked whether they want to sit with the winner or the loser of a fight. In the decisional power case, the 4- and 5-year-olds did not show any preference, but, interestingly, the 3-year-olds favoured the dominant character. In the physical supremacy case, the preference for the dominant character was higher for boys and than for girls, in line with similar findings in adults. The authors recommend however some prudence in the interpretations of these age and gender effects, since the age effect was not replicated in the physical supremacy case and the gender effect was not found in the decisional power case.

In the chapter he contributes to the collection, Alfredo Paternoster wonders "How to Be a Direct Realist". He starts from the requirement that any theory of perception should be realist in the sense that it should account for the ontological independence of the perceived object from the perceiving subject. He also hypothesizes that the view known as *direct realism*, according to which perceptual experience puts us in a direct contact with objects in the external world, is the best way to satisfy this requirement. However there are several non equivalent ways to make precise the idea that perceptual experiences put us in a *direct* contact with their object. Paternoster distinguishes relational and representational formulations of direct realism, the latter being exemplified by the views of Tyler Burge. Paternoster identifies a tension in Burge's views between his joint commitment to a principle of proximity and to externalism about content. He then argues that the only way to resolve the tension without abandoning a substantive externalism is to move towards a relational interpretation of direct realism. Thus being a relational direct realist is the best way to be a direct realist. It does not mean that this is an easy way, though. Paternoster is well aware of the difficulties that go with the relational view and points towards possible ways to solve, or at least neutralize, them.

Philip Gerrans and Kevin Mulligan close this Festschrift with a chapter entitled "Intentional Imagination and Delusion". They provide an account of imagination which integrates evidence from cognitive neuroscience and developmental psychology with philosophical arguments. According to their account, imagination is a distinctive mental process underpinned by specialized neural and computational architecture. Their intentionalist account distinguishes between the mode and content of imaginative states and proposes that the imaginative *mode* is (i) essentially independent of proximal stimulus, (ii) associative, (iii) subject to voluntary control, (iv) not ultimately responsive to the world, (v) not varying in degrees and (vi) episodic. Their account also explains the relation between complex imaginative states such as *imagining seeing, imagining hearing, imagining believing*, etc. and their counterpart states *seeing, hearing* and *believing*. I do not imagine *that* I believe that *p*, when I imagine believing that *p*. Rather, imagining-believing is the mode of my state and its content is just *p*. Imagining-believing exploits the intentional structure of belief without preserving its congruence conditions: whereas believing that *p* is congruent only if *p*, imaging-believing that *p* has no congruence conditions. Gerrans and Mulligan's account gets further support from recent work in cognitive neuroscience which identifies a specialized neural circuitry responsible for "default thinking" that meets the criteria for imagination. Their account also illuminates "doxastic borderline phenomena", i.e. imaginative states like pretense, delusion and self deception, which occupy enough of the functional role of belief that an observer might be led to think that the agent is acting on the basis of belief. Gerrans and Mulligan introduce a notion of incorporation to capture the way we act upon a mental state without metacognitively evaluating it. In cases of pretense, delusion and self-deception, agents incorporate imaginative acts to their psychology, so that these acts can play a role in the explanation of our actions. The notion of incorporation thus provides a powerful tool to integrate imagination and behaviour without either bringing imaginative states down to doxastic states, or postulating some higher-order metacognitive beliefs.

The variety of the topics dealt with in this volume hopefully gives a good, if not exhaustive, insight into the scope of Anne Reboul's research themes.[1]

References

Bjorkman B.M. (2010). "A syntactic correlate of semantic asymmetries in clausal coordination". *Proceedings of NELS* 41.
Dupuy L., A. Grabizna, N. Foudon & P. Saint-Germier (2016). *Papers dedicated to Anne Reboul.* URL: http://reboul.isc.cnrs.fr/
Korta K. & J. Perry (2011). *Critical Pragmatics*. New York: Cambridge University Press.
Levinson (2000). *Presumptive meanings: The theory of generalized conversational implicature*. Cambridge, MA: MIT press.
Moeschler J. & A. Reboul (1994). *Dictionnaire encyclopédique de pragmatique*. Paris: Le Seuil.
Noveck I.A. & A. Reboul (2008). "Experimental Pragmatics: a Gricean turn in the study of language". *Trends in Cognitive Sciences* 12(11): 425–431.
Predelli S. (2013). *Meaning without Truth*. Oxford: Oxford University Press.
Reboul A. (1986). "L'interprétation des énoncés de fiction". *Cahiers de linguistique française* 7: 27–41.
— (1987). "The relevance of *Relevance* for fiction". *Behavioral and Brain Sciences* 10: 729.

---

[1] As the editor of this volume, I would like to thank all the contributors for their participation to this project. I owe special thanks to Joanna Blochowiak for her valuable advice at the beginning of the publishing process and Jacques Moeschler for his encouragement at the very beginning of the project. I also would like to thank Ludivine Dupuy, Adrianna Grabizna and Nadège Foudon, with whom I edited the "webschrift" that started this adventure. And of course, my last words are for Anne and the inspiration her work and personality gave for all these essays.

— (1992). *Rhétorique et stylistique de la fiction*. Nancy: Presses Universitaires de Nancy.
— (2011). "Radical embodied cognition vs. 'Classical' embodied neuroscience". *The Journal of East China Normal University* 6.
— (2015a). "A new look on the Sapir-Whorf hypothesis on colours, based on neuroscientific data". In V. Bogushevskaya & E. Colla (eds). *Thinking colours. Perception, translation and representation*. Newcastle upon Tyne: Cambridge Scholars Publishing. 2–16.
— (2015b). "Why language really is not a communication system: A cognitive view of language evolution". *Frontiers in Psychology* 6: 1–12.
— (2017). *Cognition and Communication in the Evolution of Language*. Oxford: Oxford University Press.
Reboul A. & J. Moeschler (1998a). *La pragmatique aujourd'hui: une nouvelle science de la communication*. Paris: Le Seuil.
— (1998b). *Pragmatique du discours: de l'interprétation de l'énoncé à l'interprétation du discours*. Paris: Armand Colin.
Sperber D. & D. Wilson (1986). *Relevance: Communication and Cognition*. Oxford: Blackwell. (2nd edition 1995.)

Part I

Linguistics and Philosophy of Language

Chapter 2
The Semantics-Pragmatics Interface: How It Works, Why We Need It and Where It Is*

Jacques Moeschler

Abstract In this paper, three main issues are discussed: (i) How the Semantics-Pragmatics Interface (SPI) is supposed to work? (ii) Why do we need a SPI? (iii) Where is the SPI located? I show that the S-P border is porous, and that some inferred meanings are more semantic than pragmatic and vice versa. Secondly, the SPI has as a main function to allow quick and efficient information transfer, from non-linguistic source to linguistic one, and vice and versa. Finally, the SPI is mainly a linguistic issue: semantic meaning is the locus of pragmatic processes, which implies that its conceptual vs. procedural nature has some impacts on the way pragmatic meaning derivations are obtained.

Keywords Semantics-pragmatics interface · Implicature · Explicature · Accessibility.

---

*This article has been written under the SNSF research project *LogPrag: Semantics and Pragmatics of logical words* (projet n° 100012_146093) and is dedicated to Anne Reboul. Many thanks to Joanna Blochowiak and Cristina Grisot for their help and comments.

1 Introduction

During the last decades, linguistic theory has been concerned with the syntax-semantics interface, mainly with issues linked to the scope of operators (negation, quantifiers, modals) and with the syntactic or semantic nature of structural representations. One important trend in syntactic theory (for instance the cartographic approach) is devoted to the syntax-pragmatics interface, with strong arguments in favor of the syntactization of pragmatics, that is, a structural explanation of pragmatic issues, such as information structure, topic and focus and their syntactic loci in syntax (Rizzi 2013, Haegeman 2013 to cite only a few).

Even if the Semantics-Pragmatics Interface (SPI) is now in the agenda of formal semantics (Beaver et al. 2013), mainly with the aim to increase the explanatory power of dynamic semantics in accounting for context, implicature, presupposition, etc., the benefit of pragmatic theory (mainly neo- and post-Gricean approaches) has not been seriously taken into account (Moeschler 2015a).

In this article, I would like to make a series of proposals regarding the following issues:

A. How is the SPI supposed to work? Broadly speaking, is pragmatics the output of semantics or is pragmatic meaning systematically intruded onto semantics? I will show that both perspectives (pragmatics as an output and pragmatic intrusion) do not give satisfactory answer to the SPI issue. My main argument will be based on the nature of semantic and pragmatic meanings, their conventional, truth-conditional and inferential aspects. I will show that the S-P border is porous, and that some inferred meanings are more semantic than pragmatic and vice versa. The first contribution of my proposal will be that there is a continuum between semantic and pragmatic meanings.

B. Why do we need an SPI? SPI has as a main function to allow quick and efficient information transfer, from non-linguistic source to linguistic one, and vice versa. Contextual information is generally required for proposition enrichment, as well as to access contextual assumptions, in order to trigger implicit and explicit inferred meaning. On the other hand, linguistically encoded meaning is the starting point for enrichment processes in order to access reference, inferred conceptual representations, as well as implicatures (at least conventional and generalized conversational ones).

C. Where is the SPI located? The SPI is mainly a linguistic issue: semantic meaning is the locus of pragmatic processes, which implies that its conceptual or procedural nature has some impacts on the way pragmatic meaning derivations are obtained. I will give some examples of the SPI location with discourse connectives, and more precisely causal connectives.

This article is organized as follows: Section 2 explains the reason why the SPI is required in linguistic theory, and what the main proposals are since the Gricean Turn in pragmatics. Section 3 discusses the possible SPIs from a more general perspective, that is, including the relation between syntax, semantics and pragmatics. Section 4 answers the question of the function of the SPI, mainly with a discussion of scalar implicatures. Section 5 is about the location of the SPI, which will be illustrated by causal connectives, their conceptual and procedural meaning at the levels of entailment, explicature and implicature. Finally, section 6 presents a global picture of the SPI.

2 The Semantics-Pragmatics Interface

The necessity of a Semantics-Pragmatics Interface (SPI) is due to the following empirical facts: some pragmatic inferences, e.g. conversational implicatures (CI), are triggered by linguistic items (§2.1); pragmatic meaning seems to be more than non-truth-conditional, e.g. explicatures (§2.2); pragmatic meaning can be determined by truth-conditional meaning, as causal connectives show (§2.3).

2.1 Linguistic and pragmatic meanings: the case of implicatures

Generalized conversational implicatures (GCI) raise the issue of the encoding of pragmatic meaning. Are conversational implicatures (CI) attached to the semantic meaning or are they contextually triggered? The first option leads to the 'pragmatic meaning by default' solution: a CI is triggered as a default inference. On the contrary, the second option leads to the 'contextual solution': a CI must be contextually licensed or contextually blocked. For instance, how about (1) and (2)? The default solution predicts that CIs will be triggered (1-2a), whereas the contextual solution predicts that it will not (the logical reading will be inferred in (1-2b)); second, the scalar implicature (2a) in (2) is predicted, the logical reading being not accessible without a specific context (2b):

(1) Some elephants have trunks.
   a. ?? not all elephants have trunks
   b.   all elephants have trunks

(2) Some of my students passed the exam.
   a.   not all of my students passed
   b. ?? all of my students passed

So, the predictions of these two solutions are not the same. The default approach predicts that CIs should not be costly, since they are default inferences. On the other hand, the contextual approach predicts that CIs are favoured in some contexts and blocked in others. Now, experimental approaches of scalar implicatures demonstrated that the contextual approach makes better predictions than the default one (Noveck 2001, Reboul 2004, Noveck & Sperber 2007, Noveck & Reboul 2008). For instance, the logical inference (1b) is easily triggered by young children, which shows that scalar implicatures are not default inferences, but the results of the development and the maturation of a pragmatic competence.

As a consequence, the apparent advantage of the default approach – CIs are attached to lexical meaning – is ruled out by cognitive evidence. However, the contextual approach is not without disadvantages: pragmatic meaning is not calculable without accessing contextual assumptions. So although the SPI is clearly defined in the default approach, it is unclear in the contextual one. In fact, the contextual approach raises the question of what is represented in lexical meaning. To answer this question, one could use the Relevance-theoretical difference between linguistically encoded concepts and communicated inferred concepts (*ad hoc* concepts – Carston 2002, Wilson 2003, Wilson & Carston 2007). But a new question arises: what is linguistically encoded?

2.2 Explicatures vs. CIs

The second empirical fact justifying the SPI is given by pragmatic meanings that are the results of inferences and not implicit, but explicit, that is, *explicatures*. Explicatures pertain to pragmatic meaning, which is not conveyed implicitly: an explicature is an assumption that is a development of the logical form encoded by the utterance (Sperber & Wilson 1986).[1] A classical example is given by the specific meaning of *bachelor* (a young man eligible for marriage):

---

[1] "An assumption communicated by an utterance *U* is *explicit* if and only of it is a development of a logical form encoded by *U*" (Sperber & Wilson 1996: 182).

(3)   Mary is happy: she finally met a bachelor.

Whereas CIs are traditionally defined as non-truth-conditional meanings (they do not contribute to the truth-value of the proposition and they are cancellable as in (4)), explicatures are pragmatic truth-conditional meanings playing a role in the determination of the truth-value of the proposition: the truth-conditions of *P and Q* is not identical to those of *Q and P*, as (5) shows:

(4)   John fell and Mary pushed him, but not in this order.

(5)   It's always the same at parties: either I get drunk and no-one will talk to me or no-one will talk to me and I get drunk.

The consequence of the intrusion of the notion of explicature as a pragmatic meaning is evident: it reduces the area of CIs and it breaks the clear-cut border between semantics and pragmatics. This is because there are pragmatic meanings which are developments of logical forms and which are truth-conditional. Unfortunately, a new issue is raised by the notion of explicature: explicatures should not be defeasible, because this property is restricted to non-truth-conditional meaning, that is, conversational implicatures. In fact, explicatures are cancellable, as (6) and (7) show: (7) shows that the explicature of (6) [together] can be defeated without contradiction:[2]

(6)   Abi and Fée climbed the Roche de Solutré [together]

(7)   Abi and Fée climbed the Roche de Solutré, but not together.

2.3 Pragmatic meaning determined by truth-conditional meaning

Conversely, there are pragmatic meanings which are dependent on truth-conditional meanings. This is the case with the temporal and causal meanings of connectives like *and* and *because*. First, in order for *P and Q* to mean *P and then/because of this Q*, both conjuncts must be true, as (8) shows; second, in order for *P because Q* to infer that *Q CAUSE P*, both *P* and *Q* must be true (9):

(8)   #Mary pushed John and he fell, but none of these events happened.

(9)   #John fell because Mary pushed him, but none of these events happened.

---

[2]This raises the question of the criterion defining an explicature. The only possible answer is that what makes the difference between explicature and implicature lays in their truth-conditions. So, it means that the propositions expressed in (6) and (7) are not the same proposition, because the truth-conditionality property of an explicature implies that the proposition expressed and its explicatures should have the same truth-value (Moeschler 2013 for a development).

What is the empirical evidence supporting these constraints? The temporal meaning of *and* can be defeated: in this case, what is evaluated is not the truth vs. falsehood of the propositions, but the temporal relation between them (Wilson & Sperber 2012, chapter 8):

(10)   What happened was not that Peter left and Mary got angry but that Mary got angry and Peter left.

In the case of causal relations, the causal meaning of *because* cannot be defeated: what can be false is either the effect, or the causal relation: (11) can be interpreted as (12) or as (13):

(11)   John did not fall because Mary pushed him

(12)   John did not fall, and the reason is that Mary pushed him (he could fall before)

(13)   It is not because Mary pushed John that he fell, but because he slipped down.

Hence, *and* and *because* 'presuppose' the truth of the proposition they connect (Blochowiak 2014, 2016).

So, what are the provisory conclusions of this section? The first conclusion is that the SPI is more complex than the traditional Gricean pragmatics predicts. Indeed, the Gricean criteria defining the border between Semantics and Pragmatics are ruled out: (a) the truth-conditional vs. non truth-conditional aspect of meaning, (b) the cancelation criterion for implicature and (c) the implicit vs. explicit aspect of meaning.

3   Possible SPIs

What are the possible Semantics-Pragmatics Interfaces? In linguistic theory, there are at least two classical answers: (A) pragmatics as output of the linguistic system ; (B) the pervasive pragmatic intrusion into semantics. But even a superficial analysis of these solutions gives rise to negative results, because both proposals are unsatisfactory: the first solution implies a step by step processing (from syntax to pragmatics), and cannot account for pragmatic intrusion, neither for parallel processing, whereas the second solution cannot account for the relation between explicatures and implicatures, and leads to the Gricean circle. Let examine more in details these two possible, even if improbable, solutions.

3.1 The linear model

In the linear model (Moeschler & Reboul 1994, Introduction), semantics is the output of syntax, and pragmatics the output of semantics, as Figure 2.1 shows.

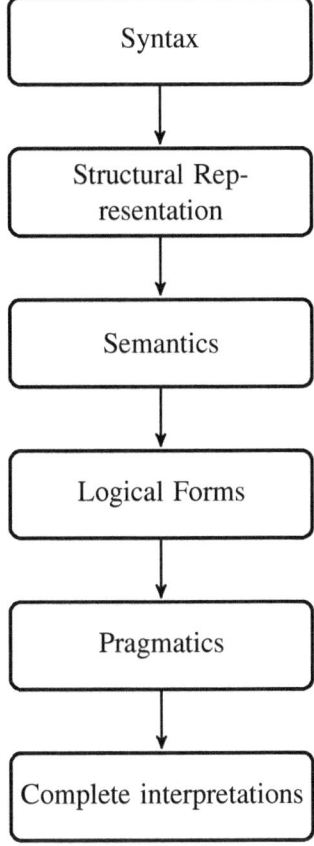

Figure 2.1: The linear model

Unfortunately, two big issues arise: first, in linguistic theory, semantics is an interface of grammar, not an output of syntax; and second, pragmatics does not belong to the linguistic system: it is not an input system (Fodor 1983), but belongs to the central system of the mind (Sperber & Wilson 1986).[3]

What does it mean for semantics to be an interface? In a formalist framework (for instance the *Minimalist Program*), logical forms (LF) are the inter-

---

[3] In the revised version of Relevance Theory (Wilson & Sperber 2012, chapter 12), there is a pragmatic module, consisting of a comprehension and an argumentative module.

face of the computational system, as phonological forms (PF) are as represented in Figure 2.2.

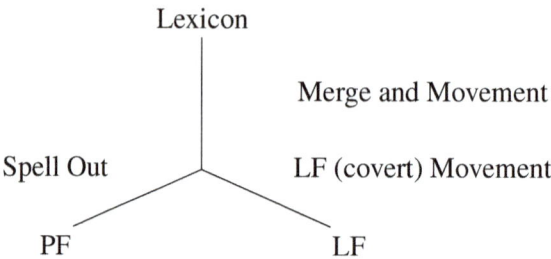

Figure 2.2: The architecture of Grammar in the Minimalist Program

In Hauser, Chomsky & Fitch (2002), interfaces are defined as the sensory-motor and the conceptual-intentional interfaces: the assumption is that phonological forms and logical forms are interfaces of the grammar, and belong to the faculty of *language in the broad sense* (FLB), whereas FLN (*faculty of language in the narrow sense*) is restricted to recursion (see Figure 2.3).[4]

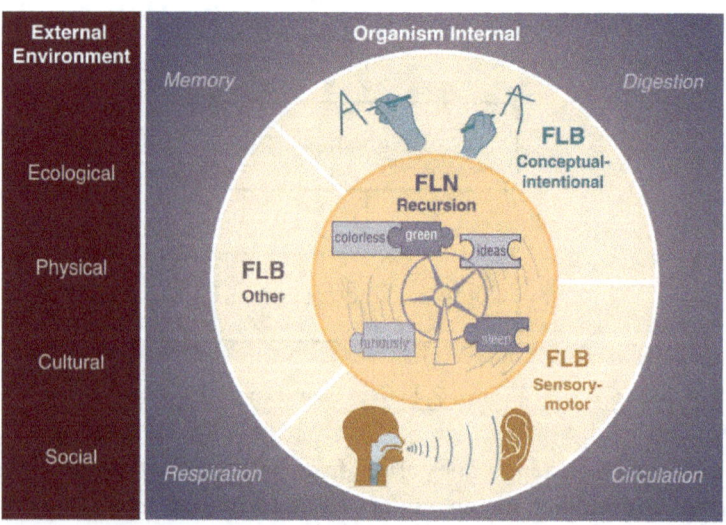

Figure 2.3: FLN, FLB and the interface of grammar (Hauser et al. 2002)

---

[4]This picture contrasts with the definition of language as form-meaning pairs, where no hierarchy between formal or semantic structures dominates (Jackendoff 2005): the flat phonological structures (PS), syntactic structures (SS) and conceptual structures (CS), implying 5 types of interfaces: interface to hearing and vocalization, PS-SS interface rules, SS-CS interfaces rules, PS-CS interfaces rules, and interfaces to perception and action.

3.2 Linguistics and pragmatics

One strong assumption of pragmatic theory is that pragmatics is not a component of linguistics, but part of the central system of thought (Sperber & Wilson 1986). In other words, pragmatic is not a module and is not devoted to specific tasks. It brings together information from different sources (linguistic, visual perception, audition, etc.). This means that pragmatics deals with different inputs (supposed to be translated into the same format) processed by the inferential central system. Linguistic information is one among other types of information processed by the central system of thought. The question that arises at this point of the discussion is: what is the relation between linguistics and pragmatics in this approach to pragmatics?

One possible answer is pragmatic intrusion. The concept of *pragmatic intrusion* implies that pragmatic interpretation affects semantic interpretation. Levinson (2000) has given number of well-known data arguing for pragmatic intrusion:[5]

a. conditional perfection (Geis & Zwicky 1971): natural language conditionals are interpreted as bi-conditionals:

(14) If you mow the lawn, I'll give you five dollars.
+> If you don't mow the lawn, I don't give you five dollars

b. conjunction buttressing (Atlas & Levinson 1981): conjunction is interpreted with more specific pragmatic meanings (temporal and causal):

(15) John turned the key and the engine started.
+> John turned the key and then/ and because of this the engine started

c. bridging (Clark & Haviland 1977): nominal anaphoras are connected with part-whole relations:

(16) John unpacked the picnic. The beer was warm.
+> The beer of the picnic

---

[5] I put aside here the many arguments given by Ross (1970) and Lakoff (1972) in favour of the Performative Hypothesis, mainly because it concerns the syntactic representation of illocutionary force, which is an issue outside the scope of what I define here as the SPI.

d. Inference on a stereotype (Atlas & Levinson 1981): stereotype information implies gender presupposed professional specialisation (a secretary is a typically a woman rather than a man):

(17)   John said 'Hello' to the secretary and then he smiled.
       +> the woman secretary

e. negative strengthening (Horn 1989): the negation of a contrary will implicate (by R/M implicature) its contrary (*not liking* weakly means *disliking*)

(18)   I don't like Alice.
       +> I dislike Alice

f. mirror maxim (Harnish 1976): in (19), the preferred interpretation is that the piano was bought by both Harry and Sue, and not that each of them bought a different piano:

(19)   Harry and Sue bought a piano.
       +> Harry and Sue bought a piano together

These facts seem at a first glance convincing: pragmatic meaning seems to interfere with semantic meaning. So what is wrong with the notion of pragmatic intrusion? The answer is straightforward: in a neo-Gricean perspective, pragmatic intrusion implies that pragmatic inferences contribute to truth-conditions. For instance, in bridging, reference resolution (as a pragmatic process) determines the truth-conditions of the full proposition. In other terms, implicatures contribute to truth-conditions, whereas the classical Gricean approach predicts that what is said contributes to what is implicated. This yields the Gricean circle (see Figure 2.4), which can be stated as follows:

(20)   The Gricean circle
       a. Implicatures (what is implicated) are computed on the basis of the proposition expressed (what is said).
       b. Implicatures determine the proposition expressed (truth-conditional meaning).

What are the repercussions of this discussion on pragmatic intrusion? First, it shows that the border between semantics and pragmatics is porous. Second, it reveals that some aspects of pragmatic meaning are truth-conditional (as

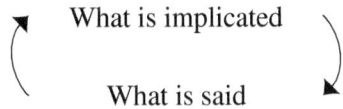

Figure 2.4: The Gricean circle

explicatures), while other are not (implicatures). And third, it becomes evident that other meaning relations need to be taken into account, as entailment and presupposition, in order to fix the SPI (Moeschler 2013).

4 The function of the SPI

Why do we need the SPI? First, the SPI has as a main function to allow quick and efficient information transfer from non-linguistic sources to linguistic ones, and vice versa. For instance, contextual information is generally required for propositional enrichment to trigger implicit and explicit inferred meaning, and it must work in a cooperative way with linguistic information. Second, even if linguistic and non-linguistic information has to be put together, the linguistically encoded meaning is the starting point of the enrichment process to access reference, inferred conceptual representations, and implicatures. In this section, I would like to show how this division of labour can be plugged in an efficient SPI by looking at the case of scalar implicatures. Scalar implicatures (SIs) are a classical case of the SPI, allowing predictions about its function.

SIs are triggered by quantifiers and are closely connected with their logical meaning, as represented by the logical square (see Figure 2.5; Horn 2004):

Horn's theory of scalar implicature connects a general principle of semantic scales and the Q-principle: a weak form implicates the negation of a strong one, the weak and strong forms belonging to the same semantic scale: so, as <I, A> and <O, E> are semantic scales, the prediction is that I implicates not-A and that O implicates not-E, as stated (21) and (22):

(21) a. I +> not-A = O
     b. O +> not-E = I

(22) a. some x +> not all x
     b. not all x +> some x

In other terms, subcontraries in the logical square implicates each other.

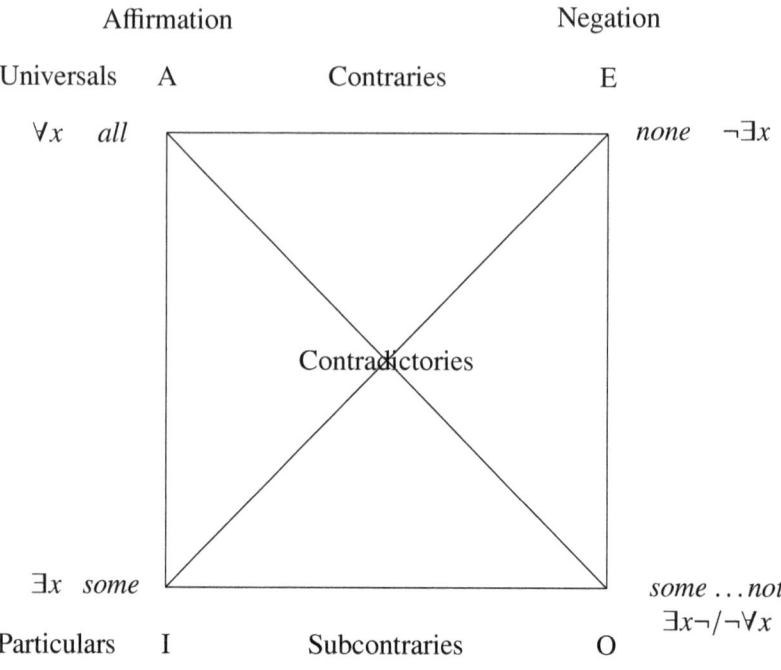

Figure 2.5: The logical square

In what follows, I will assume a strong connection between particulars, but I propose a different analysis (cf. Moeschler 2017a and 2017b for developments). I will insist on what is linguistically encoded (*semantics*) and what is inferred (*pragmatics*). The results of my analysis will be that the pragmatic meanings of *some* and *some ...not* have to be interpreted as explicatures, and not as implicatures.

Let us begin with the semantic and pragmatic meanings of particulars. What could be the semantics and pragmatics of *some* and *some ...not*? I will make here three assumptions. First, there is a strong connection between both particulars, this relation being expressed by a complement operation. Second, their semantics is defined as what is truth-conditionally incompatible with each particular: *some* is logically incompatible with *no*, as they are contradictories, and *some ...not* is logically incompatible with *all*, since they are also contradictories. Third, their pragmatics is given by their incompatibility with their upper-bound correlates: *some* is pragmatically incompatible with *all*, and *some ...not* with *none*.

So, in a nutshell, a Boolean semantics and pragmatics for *some* and *some ... not* can be given:

(23) Semantics of *some X are Y*:[6]
   a. the intersection between $[\![X]\!]$ and $[\![Y]\!]$ (the sets denoted by X and Y) is not empty;
   b. *some X are Y* is semantically incompatible with *no X is Y*;
   c. $[\![X]\!] \cap [\![Y]\!] \neq \varnothing$.

(24) Pragmatics of *some X are Y*:
   a. $[\![X]\!]$ is not included in $[\![Y]\!]$, because there must be a subset of $[\![X]\!]$ which is not in $[\![Y]\!]$;
   b. *some X are Y* is pragmatically incompatible with *all X are Y*;
   c. $[\![X]\!] \not\subset [\![Y]\!]$.

So, *some X are Y* has as pragmatic meaning its explicature *only some X are Y*. The same analysis stands for *some ... not*:

(25) Semantics of *some X are not-Y*:
   a. the intersection between $[\![X]\!]$ and the complement of $[\![Y]\!]$ (the sets denoted by X and not-Y) is not empty;
   b. *some X are not Y* is semantically incompatible with *all X are Y*;
   c. $[\![X]\!] \cap \complement([\![Y]\!]) \neq \varnothing$.

(26) Pragmatics of *some X are not Y*:
   a. the intersection between $[\![X]\!]$ and $[\![Y]\!]$ is not empty;
   b. *some X are not Y* is pragmatically incompatible with *no X is Y*;
   c. $[\![X]\!] \cap [\![Y]\!] \neq \varnothing$.

Hence, *some X are not Y* has as pragmatic meaning its explicature *only some X are not Y*.

This first analysis is not very difficult to sum up: the relation between subcontraries, that is *some* and *some ... not*, is not an implicature, but an entailment. Since their pragmatics excludes the upper-bound reading (*all* and *no*), the pragmatics of subcontraries is restricted to the truth of each of them, and not to the truth of one of them as the logical definition of subcontraries states

---

[6]This semantics is not incompatible with a proper inclusion of $[\![X]\!]$ into $[\![Y]\!]$ (its pragmatics is) or with the proper inclusion of $[\![Y]\!]$ into $[\![X]\!]$. This is the case when (i) an inclusion of $[\![X]\!]$ into $[\![Y]\!]$ is not possible, and (ii) $[\![Y]\!]$ is specifically a property attached to $[\![X]\!]$. For example, whereas *all women have children* is a false statement, *some women have children* is true, and illustrates the proper inclusion of $[\![Y]\!]$ into $[\![X]\!]$. In this case, the SI of *some* ($[\![X]\!] \not\subset [\![Y]\!]$) is blocked because of the specific semantic relation between X and Y, which satisfies the general semantics of *some* ($[\![X]\!] \cap [\![Y]\!] \neq \varnothing$).

(cf. Table 2.1 and 2.2). So each subcontrary entails the other one, since they both must be true.

P	Q	P ∨ Q
1	1	1
1	0	1
0	1	1
0	0	0

Table 2.1: The logical truth-conditions of subcontraries (inclusive disjunction)

P	Q	P ∧ Q
1	1	1
1	0	0
0	1	0
0	0	0

Table 2.2: The pragmatic truth-conditions of subcontraries (logical conjunction)

To sum up, Table 2.3 shows that the semantics *of Some X are Y* is the pragmatics of *some X are not Y*, and vice versa.

	Semantics	Pragmatics
Some X are Y	$[\![X]\!] \cap [\![Y]\!] \neq \emptyset$	$[\![X]\!] \not\subset [\![Y]\!]$
Some X are not Y	$[\![X]\!] \cap C([\![Y]\!]) \neq \emptyset$	$[\![X]\!] \cap [\![Y]\!] \neq \emptyset$

Table 2.3: the semantics and pragmatics of *some* and *some . . . not*

So, what is the difference between this analysis and the implicature analysis? The main difference lays in the truth-conditional vs. non-truth-conditional pragmatic meaning. In other words, the pragmatic meanings of subcontraries are explicatures. This raises a new question: what is the role of explicatures in utterance comprehension? Our answer is that the interpretation of particulars is directly dependent on their truth-conditional meanings, which are crucially context-dependent: the *not-all* and *not-none* interpretations can or cannot be triggered, depending on what the context is.

Now, how is the pragmatic meaning of subcontraries obtained? The assumption is that the relation with their semantics is based on an *exclusion condition*, triggering the processing of the semantics and pragmatics for *some* and *some . . . not*:

(27) The exclusion condition:
   a. exclude the incompatible semantic meaning
   b. exclude the incompatible pragmatic meaning
   c. enrich the pragmatic meaning by explicature.

In other words, this procedure yields a specification reading through narrowing the semantics of the particulars, following the heuristics given in Figure 2.6:

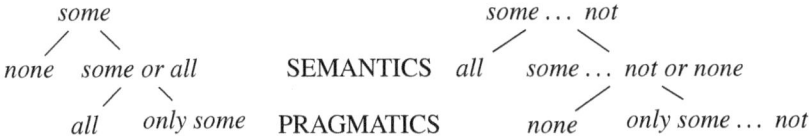

Figure 2.6: An informal heuristic for the computation of the pragmatics of *some* and *some ... not*

Figure 2.7 gives a new version of the logical square by implementing semantic and pragmatic incompatibility:

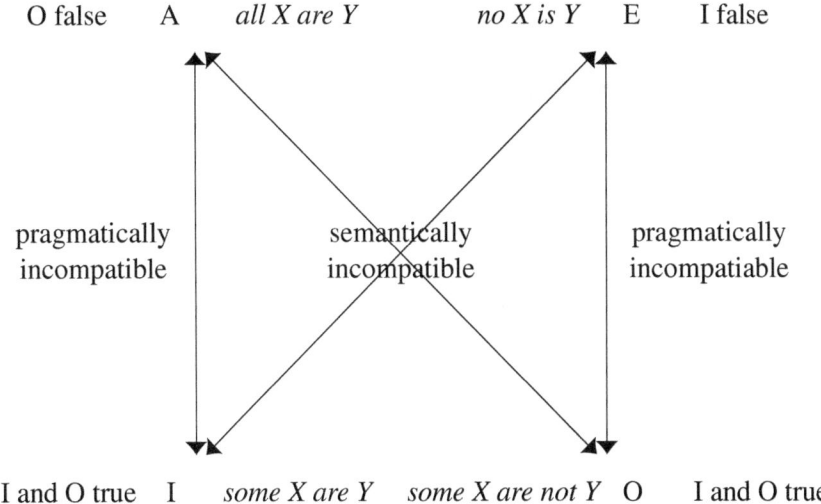

Figure 2.7: A new semantic and pragmatic logical square

We are now ready for a general explanation. Specification readings through narrowing of particulars (*only some*, *only some... not*) can receive a cognitive and communicative explanation. From a communicative point of view, two complementary explanations can be given: (i) in a Gricean perspective, it would be a violation of the first maxim of quantity in saying *some* while meaning *all*; (ii) from a Relevance-theoretical point of view, saying *some* while meaning *all* would conduct the addressee to unjustified inferences, giving rise to false conclusions, and therefore minimising the relevance of the utterance. The cognitive explanation is somehow more specific as regards the SPI: the partition reading for *some* and *some not* allows an efficient and rapid processing, avoiding useless cognitive processes. Interestingly, the prediction of this analysis is that negative particulars are not more costly cognitively than positive ones, even if they are semantically more complex (cf. Horn's conjecture[7] on negative particulars, Horn 2004 and Moeschler 2007).

5 The location of the SPI

The last question I would like to address is where is the SPI located? Let us start with the following assumption: the SPI is mainly a linguistic issue, because semantic meaning is the locus of pragmatic processes. So, in order to understand where the SPI is, we have to address the question of where conceptual and procedural meaning is located in semantic meaning. In order to answer this question, I will give some arguments from causal connectives and their semantic and pragmatic properties.[8]

How to explain the differences in semantic and pragmatic meanings between *parce que, donc,* and *et (because, therefore, and)*? Indeed, they can all have causal meanings, as in (28)-(30):

(28) Jean est tombé parce que Marie l'a poussé.
      'John fell because Mary pushed him.'

(29) Marie a poussé Jean, donc il est tombé.
      'Mary pushed John, therefore he fell.'

(30) Marie a poussé Jean, et il est tombé.
      'Mary pushed John, and he fell.

My hypothesis is that the difference is not in the meanings encoded by these connectives, but in the layers of meaning they encode. At some level, all connectives encode a CAUSE relation and allow inferring the factive vs. non-factive status of the propositions connected.

---

[7] "Given that languages tend not to lexicalize complex values that need not to be lexicalized, particularly within closed categories like quantifiers, we predict that *some ...not* will not be lexicalized, and this is precisely what we find" (Horn 2004: 11).

[8] See Moeschler (2015b) for a deeper analysis.

More precisely, in all cases, causal inferences are obtained, but with different semantic and pragmatic paths: (a) some contents are the result of entailments (Blochowiak 2010, 2014); (b) others are the result of explicatures or implicatures:[9]

(31) Jean est tombé parce que Marie l'a poussé
      'John fell because Mary pushed him'

    a. John fell & Mary pushed him

    b. Mary pushed John CAUSE John fell

(32) Marie a poussé Jean, donc il est tombé.
      'Mary pushed John, so he fell'

    a. Mary pushed John

    b. POSSIBLE (Mary pushed John CAUSE John fell)

(33) Marie a poussé Jean, et il est tombé.
      'Mary pushed John, and he fell'

    a. John fell & Mary pushed him

    b. POSSIBLE (Mary pushed John CAUSE John fell)

First, what has to be explained at the level of entailment is why *donc* does not entail the consequence (John fell), that is, the sentence it introduces. First, the truth of the consequence is not guaranteed (✗) when the cause is an event, whereas it is the case (✓) with a state (Moeschler 2011 for extended evidence):

(34) a. ✗ Marie a trop mangé, donc elle est malade.
          'Mary ate too much, so she is ill'

    b. ✗ Marie a poussé Jean, donc il est tombé.

---

[9]When a proposition is entailed, it must be true. When a proposition is developed as an explicature, it allows assigning a truth-value to the propositional form. When a proposition is an implicature, it can be cancelled.

'Mary pushed John, so he fell'

(35) a. ✓Marie est mineure, donc elle ne peut pas boire d'alcool.
'Mary is a minor, so she cannot drink alcohol'

b. ✓Axel est malade, donc le médecin le soigne.
'Axel is ill, so the doctor is treating him'

Second, the consequences in (36) (Mary is sick, John fell) are not warranted. A modal operator can be introduced in the second sentence, which shows that the consequence can be false:

(36) a. Marie a trop mangé, donc elle doit être malade.
'Mary ate too much, so she might be sick'

b. Marie a poussé Jean, donc il a dû tomber.
'Mary pushed John, so he might have fallen'

The same story works for *et*: it is compatible with situations where the cause relation is explicitly given as possible, but not certain:

(37) Marie a poussé Jean, et il est peut-être tombé.
'Mary pushed John, and he may have fallen.'

But this is not possible with *parce que*: neither the consequence (38) nor the cause (39) can be modified by a modal[10], which demonstrates the factive properties of both the cause and the consequence in the content uses of *parce que* (Sweetser 1990):

(38) #Jean est tombé parce que Marie l'a peut-être poussé.
'John fell because Mary may have pushed him.'

(39) #Jean est peut-être tombé parce que Marie l'a poussé.[11]
'John fell because Mary may have pushed him.'

Finally, with *parce que*, the causal relation can be denied:

---

[10]This is not the case with epistemic uses of *parce que* : *Jacques doit être au bureau parce que sa voiture est le parking* 'Jacques must be at work because his car is in the parking slot'.

[11]The only reading for (39) is POSSIBLE_CAUSE[Mary pushed John, John fell] and not CAUSE[Mary pushed John, POSSIBLE[John fell]] (Blochowiak 2010).

(40) Jean n'est pas tombé parce que Marie l'a poussé, mais parce qu'il a manqué une marche.
'John did not fall because Mary pushed him, but because he missed a step'.

These data support the assumption that the causal relation is a conversational implicature with *donc* and *et*, and an explicature with *parce que*.

As a summary, these three connectives trigger different degrees of speaker's commitment regarding the truth of the propositions expressed:

a. *P* is entailed by all connectives – *parce que, donc, et*.

b. *Q* is entailed by *parce que* and *et*.

c. The CAUSE relation is an explicature with *parce que*, and an implicature with iconic order under the scope of a modal operator (*et, donc*) – it can be cancelled.

Hence, the same informative content is semantically and pragmatically distributed in different ways, as Table 2.4 shows:[12]

	Entailment	Implicature	Explicature
parce que	P,Q		CAUSE (Q,P)
donc	P	POSSIBLE_CAUSE (P, Q)	
et	P,Q	POSSIBLE_CAUSE (P, Q)	

Table 2.4: Semantic and pragmatic contents of causal connectives

6 Accessibility and strength

The last issue I would like to address is the question of the impact of the type of inference on utterance interpretation. This is a relevant issue, since I made very strong proposals: SIs are not implicatures, but explicatures, and the meaning of causal connectives is shared in different layers of meaning (entailment, explicature and implicature). If these proposals make sense, then we should

---

[12] In Moeschler (2015b), I propose that entailment, explicature and implicature are conceptual meaning, distributed in semantic (entailment) and pragmatic (explicature, implicature) ones. Procedural meaning is restricted to the causal direction, iconic for *donc* and *et*, non-iconic for *parce que*.

explain why some contents are semantic and others pragmatic, and why they are distributed as they are.

The type of answer I will give to this issue is based on two concepts: *accessibility* and *strength* of meaning (Moeschler 2013). The assumption is that entailment, implicature and explicature are distributed on two scales: accessibility and strength. Accessibility defines how much a meaning is accessible to consciousness, that is, necessary to be made explicit in order to be obtained. Entailments cannot be made explicit, but pragmatic meaning as explicature and implicature can, even if explicatures are more accessible than implicatures: some implicatures are not triggered as in (1) and (2), but generally speaking, GCIs are:[13]

(41)   #I bought a Chow, so I bought a dog
       entailment: Chow(x) dog (x)[14]

(42)   Abi and Fée climbed the Roche de Solutré, and they did it together
       explicature: Abi and Fée climbed the Roche de Solutré [together]

(43)   Anne has three children, I mean no more than three.
       Implicature: Anne has no more than three children

(44) gives the accessibility scale:

(44)   Accessibility scale
       explicature > implicature > entailment

The second criterion is strength: strength defines the type of speaker's commitment. Semantic meanings, as entailment, but also presupposition[15], imply a stronger commitment than pragmatic ones, and explicatures are stronger than implicatures, because they are truth-conditional:

(45)   Strength scale
       entailment > explicature > implicature

If we put together these two scales, we obtain an interesting result:

---

[13] Here is a clear case that shows that a GCI is generally triggered: Jacques: How is my salad ? Anne: Good. Jacques: You mean, not very good? Anne: It lacks vinegar.

[14] For a general theory of conceptual hierarchy, see Reboul (2007).

[15] See Moeschler (2015a).

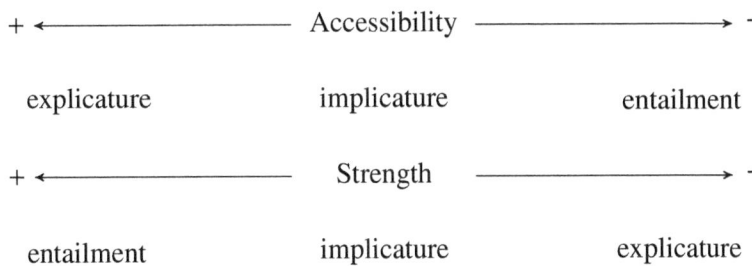

Figure 2.8: Accessibility and strength scales

Explicature and entailment are the most accessible and the strongest, entailment and implicature the less accessible and the weakest, whereas implicature and explicature are mid-ranked for accessibility and strength respectively.

What are the consequences of this picture of meaning relations? First, it shows that whatever the meaning type, lexical items are the main locus of the SPI: all my arguments have been given from lexical meaning, dispatched between semantics and pragmatics. The second consequence is that SPI, the topic of this paper, can be made visible by the continuum of semantic and pragmatic relations as entailment, explicatures and implicatures. Quantifiers as *some* and *some ... not* have shown how basic semantic relations (inclusion, intersection) are distributed at the semantic and pragmatic level, whereas causal connectives have demonstrated how propositional meanings are distributed in the semantics-pragmatics continuum.

Figure 2.9 sums up the different proposals for the location of the SPI.

7 Conclusion

In this paper, I addressed the issue of the SPI interface, and tried to give an edge to the border between semantic and pragmatic meaning. Figure 9 shows that the truly pragmatic properties (contextual, inferential accessible) are not all informative: inference is not specific to pragmatic meaning (logic is the theory of inference), context is not specific to meaning (actions have to be contextualized for instance), and that accessibility is not specific to meaning either (objects can be more or less accessible for instance). So, it means that what is more informative is not straightforward specific to semantic and pragmatic meaning: pragmatic meaning is strong and weak, explicit and implicit, truth-conditional and not-truth-conditional. Hence, the border between semantics and pragmatics definitively resembles a geographic border, shaped by landscape, instead of

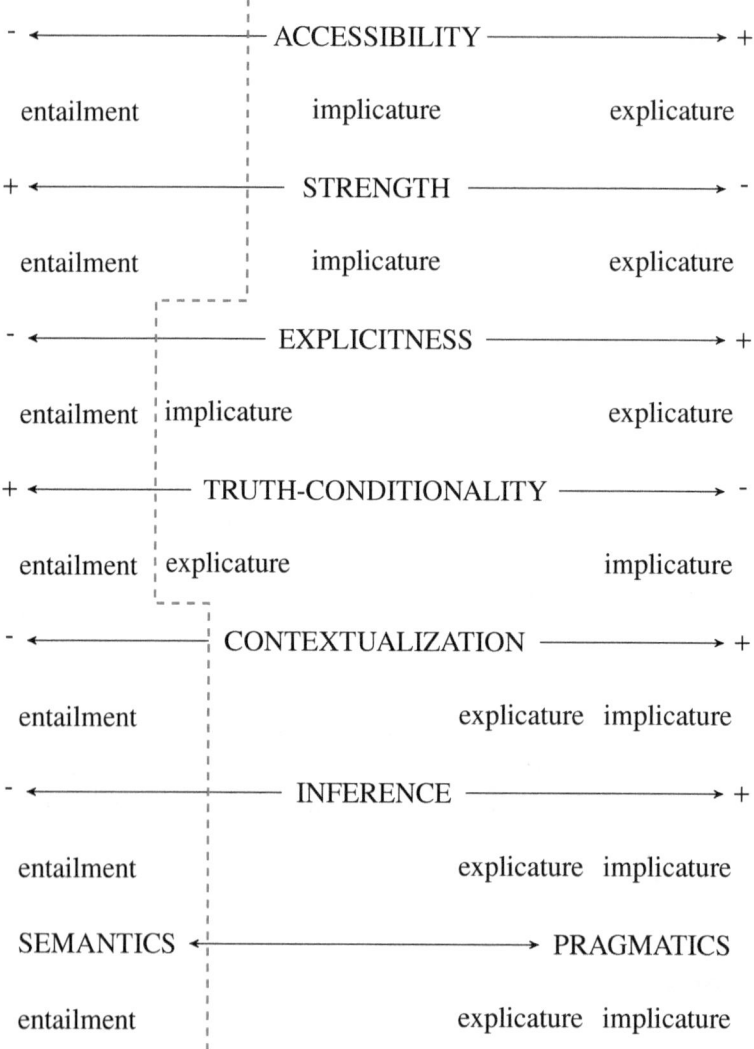

Figure 2.9: Properties of semantic and pragmatic meanings: the dotted line describes the S-P 'border'

a linear desert border.[16]

---

[16]A good geographical illustration is given by the border between two Swiss canton, Vaud and Fribourg, which includes enclaves, and mix geographical and linguistic border.

References

Atlas J. & S.C. Levinson (1981). "It-clefts, informativeness, and logical form: radical pragmatics (revised standard version)". In P. Cole (ed). *Radical Pragmatics*. New York: Academic Press. 1–61.

Beaver D., J. Cope & K. von Fitel (2013). "Semantics and pragmatics". In S.R. Anderson, J. Moeschler & F. Reboul (eds). *The Language-Cognition Interface*. Geneva: Droz. 333–351.

Blochowiak J. (2010). "Some formal properties of causal and inferential 'because' in different embedding contexts", *Generative Grammar in Geneva* 6: 191–202.

— (2014). *A theoretical approach to the quest for understanding. Semantics and pragmatics of whys and becauses*. PhD thesis, University of Geneva.

— (2016). "A presuppositional account of causal and temporal interpretations of *and*". *Topoi* 35: 93–107.

Carston R. (2002). *Thoughts and Utterances. The Pragmatics of Explicit Communication*. Oxford: Blackwell.

Clark H.H. & S.E. Haviland (1977). "Comprehension and the new-given contract". In R.O. Freedle (ed). *Discourse Production and Comprehension*. Hillsdale: Erlbaum. 1–40.

Fodor J.A. (1983). *The Modularity of the Mind*. Cambridge: MIT Press.

Geis M. & A. Zwicky (1971). "On invited inferences". *Linguistic Inquiry* 2: 561–566.

Haegeman L. (2013). "The syntax of adverbial clauses". In S.R. Anderson S.R., J. Moeschler & F. Reboul (eds.), *The Language-Cognition Interface*. Geneva: Droz. 135–156.

Harnish R. (1976). "Logical Form and Implicature". In T. Bever, J. Katz & T. Langendoen (eds). *An Integrated Theory of Linguistic Abilities*. New York: Crowell. 464–479.

Hauser M., N. Chomsky & W.T. Fitch (2002). "The faculty of language. What is it, who has it, and how did it evolve?". *Science* 298: 1569–1579.

Horn L.R. (1989). *A Natural History of Negation*. Chicago: The Chicago University Press.

— (2004). "Implicature". In L.R. Horn & G. Ward (eds). *The Handbook of Pragmatics*. Oxford: Blackwell: 3–28.

Jackendoff R. (2005). *Foundations of Language. Brain, Meaning, Grammar, Evolution*. Oxford: Oxford University Press.

Lakoff G. (1972). "Linguistics and natural logic". In D. Davidson & G. Harman (eds). *Semantics of Natural Language*. Dordrecht: Reidel. 545–665.

Levinson S.C. (2000). *Presumptive Meanings. The Theory of Generalized Conversational Implicatures*. Cambridge: MIT Press.

Moeschler J. (2007). "Why are there no negative particulars? Horn's conjecture revisited", *Generative Grammar in Geneva* 5: 1–13.

— (2011). "Causal, inferential and temporal connectives: Why *parce que* Is The only causal connective in French". In S. Hancil (ed). *Marqueurs discursifs et subjectivité*. Rouen: Presses Universitaires de Rouen et du Havre. 97–114.

— (2013). "Is a speaker-based pragmatics possible? Or how can a hearer infer a speaker's commitment?". *Journal of Pragmatics* 43: 84–97

— (2015a). "La frontière sémantique-pragmatique existe-t-elle? La question des présuppositions et des implicatures révisitée". In A. Rabatel, A. Ferrara-Léturgie & A. Léturgie (eds). *La sémantique et ses interfaces*. Limoges: Ed. Lambert-Lucas. 263–288.

— (2015b). "Where is procedural meaning? Evidence from discourse connectives and tenses". *Lingua* doi:10.1016/j.lingua.2015.11.006

— (2017a). "Back to negative particulars. A truth-conditional account". In S. Assimakopoulos (ed) *Pragmatics and its Interfaces*. Berlin: Mouton de Gruyter. 7–32

— (2017b). "Formal and natural languages: what does logic tell us about natural language?". In A. Barron, G. Steen & Y. Gueguo (eds). *The Routledge Handbook of Pragmatics*. London: Routledge. 241–256.

Moeschler J. & A. Reboul (1994). *Dictionnaire encyclopédique de pragmatique*. Paris: Ed. du Seuil.

Noveck I. (2001). "When children are more logical than adults: investigations of scalar implicature". *Cognition* 78(2): 165–188.

Noveck I. & A. Reboul (2008). "Experimental pragmatics: A Gricean turn in the study of language", *Trends in Cognitive Sciences* 12(11): 425–431.

Noveck I. & D. Sperber (2007). "The why and how of experimental pragmatics". In N. Burton-Roberts (ed). *Pragmatics*. Basingstoke: Palgrave McMillan. 141–171.

Sperber D. & D. Wilson (1986). *Relevance. Communication and Cognition*. Oxford: Blackwell.

Reboul A. (2004). "Conversational implicatures: nonce or generalized?". In I.A. Noveck & D. Sperber (eds). *Experimental Pragmatics*. Basingstoke: Palgrave McMillan. 322–333.

— (2007). *Langage et cognition humaine*. Grenoble: Presses Universitaires de Grenoble.

Rizzi L. (2013). "Theoretical and comparative syntax: some current issues". In S.R. Anderson, J. Moeschler & F. Reboul (eds). *The Language-Cognition Interface*. Geneva: Droz. 307–331.

Ross J.R. (1970). "On declarative sentences". In R.A. Jacob & P.S. Rosenbaum (eds). *Readings in English Transformational Grammar*. Waltham: Ginn. 222–272.

Sweetser E. (1990). *From Etymology to Pragmatics*. Cambridge: Cambridge University Press.

Wilson D. (2003). "Relevance and lexical pragmatics". *Italian Journal of Linguistics* 15(2): 273–291.

Wilson D. & R. Carston (2007). "A unitary approach to lexical pragmatics: relevance, inference and ad hoc concepts". In N. Burton-Roberts (ed). *Pragmatics*. Basingstoke: Palgrave McMillan. 230–259.

Wilson D. & D. Sperber (2012), *Meaning and Relevance*. Cambridge: Cambridge University Press.

Chapter 3
The Semantics of Functional Prepositions in Serbian

Tijana Ašić

Abstract Some prepositions in Serbian denote the existence of an active relationship between the figure and the ground and communicate that the subject is activating the function of the prepositional argument. However, the agentivity always interferes with the specific position of a figure in relation to the ground. In this paper we test whether the semantics of the prepositions in question can be defined using spatial predicates only or if it has to be based on some additional functional concepts.

Keywords Functional prepositions · Serbian · Mereotopology · Figure · Ground

1 Introduction

In our previous work on spatial, spatiotemporal and temporal prepositions (Ašić 2008, Ašić and Stanojević 2013) we have shown that it is possible to define them using a non *ad hoc* and general, spatial ontology, such as the one created by Casati and Varzi in their book (1999). Moreover, we have created definitions of the prepositions in question consisting of basic mereo-topological predicates (*contact, connection, inclusion*) augmented by the directional notions (computed by the frames of reference; see Levinson 2003).[1] In this per-

---

[1] A frame of reference is a coordinate system used to identify location of an object. In languages different frames of reference can be used.

spective none of the functional relations between a figure and a ground is a part of the semantics of preposition but it is a result of pragmatic inferential process.[2] Our approach obeys Grice's well-known principle of modified Occam's razor (Grice 1978) which enjoins semanticists not to multiply senses beyond necessity.

For example, the preposition *na (on)* can be defined with a predicate called external connection (defined in Casati and Varzi's spatial ontology):[3]

Definition 1 (External connection). $ECxy =_{df} Cxy \land \neg Oxy$
("x is externally connected with y" is equal by definition to "x is connected with y and x does not overlap y".)

Definition 2 (*na*). $xnay =_{df} ECxy$
("x is na y" is equal by definition to "x is externally connected with y".)

As for the "carrier/carried" relation (often, but not always existing in this relation), it does not exist in its definition and has to be pragmatically inferred.

As for the prepositions based on the notion of direction, such as *ispred / iza (in front of / behind)* they can be defined with notions such as *positive / negative frontal region* (which depends on the chosen reference frame): *Ispred* designates either that the figure is situated on the frontal side of the ground (intrinsic frame of reference) or that the figure is closer to the observer / some other reference point (relative frame of reference).

Definition 3 (Positive frontal region).
$pfr(x) =_{df} \imath w(Rw \land Clw \land Pw(n(x)) \land \neg Pxw \land FPw(r(x)))$
(The positive frontal region to x is equal by definition to the unique closed region w that is part of the neighborhood of x, without x being part of it, and that it is frontally positive to the region of x.)

Definition 4 (Negative frontal region).
$nfr(x) =_{df} \imath w(Rw \land Clw \land Pw(n(x)) \land \neg Pxw \land FNw(r(x)))$
(The negative frontal region to x is equal by definition to the unique closed region w that is a part of the neighborhood of x, without x being part of it, and that is frontally negative to the region of x.)

---

[2] A figure is a moving or conceptually movable entity whose site, path or orientation is conceived as a variable, the particular value of which is the relevant issue. A ground is a reference entity, one that has a stationary setting relative to a reference frame.

[3] The mereo-topological predicates used in the following definitions follow the notational conventions of (Casati and Varzi 1999). This ontology deals with spatial objects and relations between them.

Definition 5 (*ispred*). x*ispred*y $=_{df} \exists w(Pw(pfr(y)) \wedge RLxw)$

("x is ispred *y" is equal by definition to "there is some w such that w is a part of the positive frontal region of y and x is exactly co-located with w")*

Definition 6 (*iza*). x*iza*y $=_{df} \exists w(Pw(nfr(y)) \wedge RLxw)$

("x is iza *y" is equal by definition to "there is some w such that w is a part of the negative frontal region of y and x is exactly co-located with w")*

However, further investigations of the spatial expressions in Serbian (Ašić 2005, 2006.) have shown that definitions of some semantically complex prepositions should be enriched by specifications of the constraints they impose on the nature of the figure and the ground. For example, the preposition *po (over)* just like *na,* denotes the relation of contact but also demands that the figure is either continuous by nature or that it is moving on the ground (Ašić 2005).[4] Likewise, the preposition *uz* puts strong constraints on the dimension and shape of the figure and the ground (Ašić 2006).

However, advocates of the functional approach state that some spatial prepositions do not only denote the position of the figure in relation to the ground, but also a type of physical relation between them (Klikovac 2004). Hence, their definitions should consist of functional notions such as *force, support, conflict, action, intention, manipulation* etc. Typical examples of "functional spatial prepositions" would be *pred* and *za.* Namely, *biti pred televizorom* (to be *pred* TV) means *to watch TV,* and *biti za volanom* (to be *za* steering*)* means *to drive.*

Before we check the validity of the functionalist approach, we will make an overview of all the different usages of these two prepositions.

2 The preposition *pred*

2.1 The meaning of *pred*

In addition to the spatial directional preposition *ispred,* there is in Serbian also a preposition *pred,* which is morphologically related to it[5] (see also Ašić and Stanojevic 2008). Typical examples of the usage of *pred* suggest that it is used when a speaker wants to denote not only a spatial relation between a figure and a ground, but also the fact that the figure (usually an animate entity) is

---

[4]*Knjiga je na stolu* / A book is on the table. *Brasno je svuda po stolu* / The flour is all over the table. *Mis trci po stolu* / A mouse is running on the table.

[5]They both contain the same root *pred* which has a *frontal* meaning.

activating the ground's function.[6] Note that the ground is usually an object which would be called the prominent telic quale[7] in the Generative lexicon (see Pustejovsky 1995).

(1) *Dušan je pred televizorom. Gleda crtani.*
     Dušan is in front of the TV set. He is watching a cartoon.

(2) *Pas je pred vratima. Pokušava da ih otvori.*
     The dog is in front of the door. It is trying to open it.

(3) *Dušan je pred Moneoovom slikom. Divi se nijansama narandzaste boje.*
     Dušan is in front of a Monet's painting. He is admiring nuances of the orange colour.

Contrasting the following two examples may serve as a strong argument for a qualification of the preposition *pred* as *functional:*

(4) *Dečak stoji ispred automata za sokove i čeka.*
     A boy is standing in front of the beverage wending machine and waiting.

(5) *Dečak stoji pred automatom za sokove i čeka.*
     A boy is standing before the beverage wending machine and waiting.

While the first example conveys the information about the subject's position without specifying the object of his waiting, the second one communicates that he is waiting for a beverage from the wending machine. The crucial question is whether this meaning is yielded by the semantics of the preposition *pred* which denotes an active relation between the figure and the ground or if the supposed usage of the wending machine is a pragmatic implicature.

One thing is certain: with *pred*, the orientation is encoded exclusively by the intrinsic frame of reference (while with *ispred*, it can also depend on the relative frame of reference) which is relevant for both entities in the relation: the figure is situated on the frontal side of the ground, but it also faces the ground with its frontal part.[8] This is why the following example is unaccept-

---

[6]On the activation of the gound's function with the prepositions *na* and *u* in the telic constructions, see Ašić and Corblin (2014).

[7]This means that there is a purpose for which it was made or for which it is to be used. For example the telic quale of the book is *to be read.*

[8]The intrinsic frame of reference is based on the inherent spatial features of objects (their frontal, back or lateral parts, etc.). The relative frame of reference is based on the observer's view-point.

able:

(6)   *Dušan je pred ogledalom ali mu je okrenut ledjima[9].
      Dušan is in front of the mirror, but his back is turned towards it.

Does it follow from the statement given above that *pred* is actually a symmetric preposition, just like *en face* in French? A strong argument against this position resides in the fact that the inversion usually leads to a sentence with a slightly different meaning and in some cases it is clearly impossible:

(7)   Dušan je pred ogledalom.
      Dušan is in front of the mirror.
(7') *Ogledalo je pred Dušanom.
      The mirror is in front of Dušan.

Does this suggest that the "facing" relation is not sufficient for defining the meaning of *pred*? Is a functionality really a part of the basic semantics of this preposition? If this is so then it would be impossible to cancel it without making the sentence false (see Stvan 1998 for a discussion on implicatures created by prepositions). However, the following example shows that this is possible:

(8)   Dušan je pred ogledalom ali se ne ogleda. Oci su mu zatvorene.
      Dušan is in front of the mirror but he's not looking at himself. His eyes are closed.

We have thus proved that the functionalist definition is not adequate for *pred*; it seems, therefore that the definition of this preposition should be based on the notion of the frontal orientation.

However, some examples present a problem for such a definition. Actually we have to account for the non-symmetrical nature of this preposition and also for the possibility to use it with objects with no intrinsic frontal part (this is true both for a figure and a ground).

(9)   Trkač je pred linijom cilja.
      The runner is approaching the finish line.
(10) Lopta je pred golom.
      The ball is in front of the goal.

---

[9]Remember that this sentence becomes acceptable if we replace *pred* with *ispred*: *Dušan je ispred ogledala ali mu je okrenut ledjima* / Dušan is in front of the mirror, but his back is turned towards it.

What motivates the usage of *pred* in these sentences? Although the figure is used with a stative verb, there is an understatement of motion: the figure is captured *on its way* to the ground. It is no wonder that this preposition (and not *ispred*) is commonly used to express the vicinity to the final point of a trajectory:

(11)   *Pred Beogradom smo. Vidim Avalski toranj.*
        We are approaching Belgrade. I can see the Avala Tower.

In accordance with the Localistic Hypothesis (according to which all the expressions used to denote abstract relations have a basic spatial meaning), the spatial proximity easily becomes a temporal proximity. Thus, the following sentence is ambiguous:

(12)   *Gosti su ti pred vratima.*
        Your guests are in front of the door.
        Your guests have almost arrived.

Our next task, then, should be to relate the mentioned dynamic and static usages of *pred*. Our suggestion is that even the usages where there is no movement in the direction of the ground can be considered as fictionally dynamic. In the case of watching TV or observing a painting or looking at oneself in the mirror, the sight is directed towards the ground ("eyes are travelling towards the ground"). In a sense, the figure is mentally approaching the ground.[10] This dynamic constraint on the figure explains why the preposition *pred* is not symmetric like *en face de* and why the sentence in which we invert them are often unnatural[11].

Interestingly, this preposition can be used even with a motionless figure with no intrinsic frontal part. The condition for this usage is that the picture captures also a movement of a third object: the figure is situated on its itinerary to the ground:

(13)   *Deca trče ka ulazu u školu. Izgleda kao da će svi ući na vreme. Ali ne! Bara je pred školom. Deca moraju da je obidju. vrata se zatvaraju. Ostali su napolju!*
        The children are running towards the school. It seems that they will get there on time. But no! There is a puddle in front of / before the school. The children have to bypass it. The door is closing. They have remained outside.

---

[10] See Talmy (2000) for an account of a fictive motion.

[11] *Televizor je pred Dušanom* / Tv is in front of Dušan.

If there is no indication of such a movement the sentence is unacceptable.

2.2 *Pred* with animated ground

The sentences in which both the figure and the ground of the preposition *pred* are animated (usually human beings) have a specific meaning. They imply the existence of a visual contact in both directions.

(14) *Klovn je izveo skeč pred decom*
      The clown performed a skit in front of the children.

This sentence communicates not only the position of the subject, but also the fact that the ground is watching and consequently can affect him.

Quite often the preposition *pred* denotes the presence of the figure in the visual field of the ground. In this usage the actual position of the figure is irrelevant.

(15) *On ne puši pred roditeljima, jer su jako strogi.*
      He does not smoke in front of his parents because they are very strict.

The implication of "being observed" also accounts for the fact that some authors analyse it as a causal preposition; namely, it appears with verbs denoting psychological reactions triggered by the ground's presence. As they point out, the causal force is triggered by the visual contact with the ground (see Piper 2005).

All in all, *pred* does not serve to convey the exact position of the figure, but to point to the dynamic relation between its arguments. This characteristic prevents its modification with quantification phrases. Contrary to this, *ispred* is often modified by this type of adverbs, but *pred* cannot be so modified:

(16) a. *Kornjača je na početku trke 10 metara' mnogo ispred zeca.*
          At the beginning of the race the turtle is 10 metres – very much in front of the rabbit.
      b. **Kornjača je na početku trke 10 metara' mnogo *pred zecom.*
          At the beginning of the race the turtle is 10 metres – very much in front of the rabbit.

(17) a. *Dva metra ispred komore nalazi se sveća sa plamenom visokim 3 cm.*
          Two metres in front of the chamber there is a candle with a three-centimeter high flame.

b. *Dva metra pred komorom nalazi se sveća sa plamenom visokim 3 cm.
Two metres in front of the chamber there is a candle with a three-centimeter high flame.

2.3 Abstract usages of *pred*

The basic semantics of *pred* based on the notion of the figure moving towards the ground can also explain its abstract usages. In accordance with the Localistic Hypothesis, it is used to denote temporal anterior proximity. The metaphor we have is that the figure-event is situated just before the ground-event (Piper 2001, 138).

(18)   *Došli su minut pred ručak.*[12]
       They arrived just before lunch.

(19)   *Daće intervju neposredno pred početak nove sezone u formuli 1.*
       He will give an interview just before the start of the new Formula One season.

As we can see from the examples, temporal *pred* can be modified by adverbials denoting a very short temporal distance, but not with adverbials denoting a bigger temporal distance:

(20)   a.   *Došli su tri sata pre ručka.*
            They arrived three hours before lunch.
       b.   **Došli su tri sata pred ručka.*
            They arrived three hours before lunch.

In this case the neutral temporal preposition *pre (before)* has to be used.

2.4 *Pred* with events

Let us finally examine a very interesting case in which the preposition *pred* is used with some abstract entities, namely events. Its function is not only to show that the event in question will happen in the near future; it also serves to show that the subject has to deal with it or to overcome it:

(21)   *Oni su pred razvodom.*
       They are about to divorce.

---

[12] There is a change in case—instead of the instrumental case, we have the accusative case here, but this point will not be discussed in this paper.

(22) *Oni su pred selidbom u novu kucu.*
 They are preparing for moving into a new house.

(23) *Francuska je u tom trenutku pred revolucijom.*
 France was on the eve of the Revolution at the time.

(24) *On je pred velikom odlukom.*
 He is now faced with a major decision/ A major decision lies ahead of him.

(25) *Predsednik tvrdi da smo pred rešenjem tog problema.*
 The President claims we are about to find the answer to that problem.

(26) *Novine pišu da je Bašar al-Aasad pred porazom.*
 The newspapers report Bashar al-Aasad faces imminent defeat.

Our assumption is that this usage of *pred* is conceptually related to the usage in which the figure and the ground are facing each other. In other words, the animate figure has to confront the ground. This imposes a constraint on the nature of the ground: it has to be an important event, an entity which is, metaphorically, powerful. The subject has to invest some energy to reach it.

If the event lacks this power, the sentence becomes unacceptable:

(27) *Milica je pred rođendanom.*
 *Milica is now faced with her birthday.

3 The preposition *za*

Among many usages of the preposition *za* in Serbian there is one that seems to denote not only the contact but the functional relation between the figure and the ground. More precisely, as shown in the following example, it denotes that the figure (a human being) is performing an activity using the ground (obligatory inanimate object with a specific function):

(28) *Ema je za klavirom. Divno svira.*
 Emma is at the piano. She is playing in a wonderful way.

However, just like with *pred* it is possible here to cancel the activity implicature:

(29) *Ema je za klavirom ali ne svira, već briše dirke.*
 Emma is at the piano, but she is not playing—she is cleaning the keys.

Note that it is not possible to cancel the figure's position in relation to the ground:

(30)   *Ema je za klavirom ali mu sedi okrenuta ledjima.
       Emma is in front of the piano, sitting with her back turned to it.

It means that there is a strong constraint on the type of the position of the figure related to the ground. If we replace *za* with *na* (designating a weak contact between a figure and a ground) we get a mere spatial reading:

(31)   Ema je na klaviru. Pašće!
       Emma is on the piano. She is going to fall!

However, the definition of the basic semantics of *za* as a specific contact cannot explain all the different usages of this preposition.

It should be noted that *za* is actually the shorter form of the basic spatial preposition *iza* denoting that the figure is situated in the negative frontal region of the ground. *Za* in its basic spatial usage chooses a moving figure which is getting closer to the ground. The image we have is that the figure is trying to catch the ground:

(32)   Dečak trci za devojčicom.
       A boy is running after a girl.

Note that if we replace *za* in this sentence with *iza*, we get a different image: a boy is just running behind the girl.

(33)   Dečak trci iza devojčice.
       A boy is running behind a girl.

This preposition can be used with a static predicate on the condition that the ground is an entity with salient telic qualia. A dynamic feature "approaching the ground" transforms itself into the "intention to use the ground".

4   Conclusion

Our analysis has shown that the definition of the so called "functional preposition" can be based exclusively on spatial predicates and specific constraints on the nature and dynamicity of the figure and the ground.

Our observations on the semantics of *pred* and *za* present quite a strong argument for the Aurnague, Vieu Borillo (1997: 24) three-level theory that enables the proper representation of prepositional meaning. The geometric level forms a basis of this system. The definitions of prepositions are generated at this level.

The functional level captures the features of the figure and the ground and the non-geometric relations between them. It actually concerns our representation of the typical position of the figure related to the ground.

Finally, the pragmatic level is based on the extra-linguistic information, such as context. Thanks to this enrichment we are able to understand that in the example (5) the focus is not on the subject's position but on his activity (waiting for a beverage).

Our findings suggest that the difference between the spatial and spatio-functional prepositions should not be looked for at the basic geometric level, but at the second one.

References

Ašić T. (2005). "The po-na-u opposition in Serbian an its Equivalent in Bulgarian". *Balkanistica. A Journal of Southeast European Studies* 18: 1–30.
— (2006). "Les usages temporels de la prépositions uz en serbe est leurs équivalents en français". *Nouveaux cahiers de linguistique française* 27: 50–65.
— (2008). *Espace, temps, préposition*. Geneva: Droz.
Ašić T. & V. Stanojević (2008). "O predlozima ispred i pred u srpskom jeziku". In M.Radovanović & P. Piper (eds). *Semantička proučavanja srpskog jezika*. Belgrade: SANU. 129–150.
Ašić T. & F. Corblin (2014). "Telic definites and their prepositions: French and Serbian". In A. Aguilar-Guevara, B. Le Bruyn & J. Zwarts (eds). *Weak Referentiality*. Amsterdam: John Benjamins. 183–212.
Aurnague M., L. Vieu, A. Borillo (1997). "Représentation formelle des concepts spatiaux dans la langue". In M. Denis (ed). *Langage et cognition spatiale*. Paris: Masson. 69–102.
Casati R. & A. Varzi (1999). *Parts and places, The Structures of Spatial Representation*. Cambridge MA: MIT press.
Grice H.P. (1978). "Further Notes on Logic and Conversation," in P. Cole (ed). *Syntax and Semantics: Pragmatics* 9. New York: Academic Press. 183–97.
Klikovac D. (2004). *Semantika predloga - Studija iz kognitivne lingvistike*. Belgrade: Filosoški fakultet.
Levinson S.C. (2003). *Space in Language and Cognition*. Oxford: Oxford University Press.

Piper P. (2001). *Jezik i prostor*. Belgrade: XX vek.
— (2005). *Sintaksa srpskog jezika, prosta rečenica*. Belgrade: SANU.
Pustejovsky J. (1995). *The Generative Lexicon*. Cambridge MA: MIT Press.
Stvan L. (1998). *The semantics and pragmatics of bare singular noun phrases*. PhD Dissertation, Northwestern University.
Talmy L. (2000). *Toward a Cognitive Semantics*. Cambridge MA: MIT Press.

Chapter 4

De Se Readings as a Window on the Nature of Control

Denis Delfitto & Gaetano Fiorin

Abstract This paper argues that a specific (sub-)class of *de se* readings found in certain Control structures and in a well-defined set of lexical reflexives provides the key for the understanding of some long-debated and presently still poorly understood properties of Control. We show that the phenomenon known in the philosophical literature as 'immunity to error through misidentification' offers an important window on the nature of Control and on its core interpretive properties, including the distinction between Exhaustive and Partial Control.

Keywords Control · Immunity to error through misidentification · Theta-roles · Attitudes *de se*.

1 Introduction

This paper argues that a specific class of *de se* readings found in Control structures and, arguably, also in structures involving lexical reflexives provides the key for the understanding of some long-debated and presently still poorly understood properties of Control. In a nutshell, we will argue that the contrast between the semantics of a sentence like "I visited East-Berlin 30 years ago", whereby it makes sense to ask "Is it really you who visited E-B 30 years ago?" and a sentence like "I am in pain", for which the equivalent questioning (*Is it really you who is in pain?*) would be meaningless[1] extends to complex structures involving Control. In fact, if I utter the sentence "I think that I have visited E-B 30 years ago", intending to the tell the truth, it might still be meaningful for you to express doubts (say, based on a different recollection of the relevant events) on the identity of the subject of the embedded clause, by asking for instance: "Is it really you the person of whom you think that she visited E-B 30 years ago?". Conversely, if I utter *bona fide* "I remember visiting E-B 30 years ago", and I intend my speech act as a report on the phenomenal experience of which I am presently the bearer, it sounds pointless for you to inquire about the referential identity of the embedded subject by asking "Is it really you the person of whom you remember that she visited E-B 30 years ago?". We propose that this phenomenon (known in the philosophical literature as 'immunity to error through misidentification', from now on IEM) offers an important window on the nature of Control and on its core interpretive properties, including the distinction between Exhaustive and Partial Control (from now on EC and PC, respectively).

The paper is organized as follows. In section 2, we discuss the relation between Control and restructuring, based on Cinque's insight[2] that the cases where Control can be reduced to Raising (i.e. to A-movement, as in the Movement Theory of Control (MTC))[3], overlap with the cases where Control is interpreted exhaustively, whereas the residual cases, in which Control involves, by hypothesis, the syntactic realization of the embedded subject as an empty category (PRO), overlap with the cases of Partial Control (or 'imperfect control', as the phenomenon is also dubbed). We will argue that Cinque's analysis, though conceptually elegant, is empirically untenable, since (as discussed in detail in section 5) there are restructuring verbs (like *volere*/want) that readily admit Partial Control, and, on the other side, non-restructuring verbs that do not admit Partial Control (like believe/*credere*[4]).

In section 3 we present our solution to what we dub *Cinque's paradox* (i.e. 'How can restructuring verbs be allowed to discharge a theta-role in syntax?') by proposing that the theta-role assigned by the higher predicate (i.e. the experiencer) overwrites the theta-role associated with the lower predicate (i.e. the agent). This approach elegantly derives the fact that restructuring verbs are strictly bound to the specific class of *de se* readings that give rise to IEM-effects.

---

[1] See Wittgenstein (1958), Shoemaker (1968), Pryor (1999), Recanati (2007) among others.
[2] See Cinque (2004).
[3] See Hornstein (1999) and subsequent work by the same author.
[4] See Pearson (2013).

In section 4, we present additional empirical evidence for the claim that thematic overwriting is a UG option whenever two theta-roles turn out to be semantically indistinguishable on metaphysical grounds, by providing an analysis of lexical reflexives that explains why a subclass of these predicates gives rise to IEM-effects.

Section 5 shows that IEM-effects extend beyond the class of restructuring predicates. In turn, this observation can be taken to show that at least a subclass of non-restructuring control predicates is incompatible with the syntactic realization of the embedded subject (say, as PRO). We will further explore whether there are empirical and conceptual grounds to adopt a weaker version of Cinque's hypothesis on the distribution and interpretation of PRO, according to which all control structures that do not give rise to IEM-effects require the syntactic realization of PRO as the embedded subject.

In section 6, we develop a principled solution for the many puzzles arising with respect to Partial Control. First, we observe that PC is found even with non-restructuring IEM-predicates, and this may be taken to show that PC cannot be a function of the semantics of PRO (since, by hypothesis, there is no PRO in control structures featuring IEM-effects). Second, we argue that the correct distribution of PC across control predicates can be derived as a combined effect of the semantics of the higher predicate and the semantics of thematic overwriting. More particularly, we show that PC is correctly predicted to arise, as a pragmatic effect, under the view that thematic overwriting makes the overwritten theta-role interpretively available in the pragmatic component, essentially through pragmatic enrichment.

In the final section, we draw some conclusions, emphasizing that our approach, based on thematic overwriting, offers an elegant analysis of the most puzzling interpretive properties of Control. We also suggest that MTC is not viable as the syntax of control but remains the correct syntax for restructuring control predicates, as proposed by Cinque. At the same time, the semantics of PC does not depend on the presence of PRO since it extends to restructuring predicates displaying IEM-effects. In our analysis, the right results follow from an optimal balance of syntactic and interpretive ingredients, with virtually no recourse to empirically unmotivated stipulations.

2 Restructuring and control

Cinque (2004) offers a principled analysis of restructuring predicates, based on the insight that these are not lexical verbs projecting an independent argument structure but rather functional elements directly inserted into a head position within the articulated functional hierarchy proper to the clause. There is no need to hypothesize overt or abstract restructuring operations yielding monoclausality as an effect of verb-movement, as in Rizzi's original analysis[5] and much of subsequent work on the topic. In fact, monoclausality results from the very nature of the 'restructuring' predicate as a functional element. In favor of this analysis Cinque adduces the rigid relative order among restructuring verbs that is observed in languages like Italian, and the fact that monoclausality does not seem to be rigidly tied to the manifestation of canonical 'transparency effects' such as clitic-climbing. More particularly, Cinque argues that there is no solid empirical evidence in favor of the idea that restructuring involves complex verb formation through head-movement. Rizzi's traditional syntactic tests in favor of this kind of syntactic constituency (such as Right Node Raising and Cleft Sentence Formation) are critically discussed and dismissed, under the claim that other syntactic operations (Like Focus Movement and Topicalization) are admitted to apply to structures featuring clitic-climbing. From the present paper's perspective, an important result emerges from this analysis: for the subset of control structures that admit restructuring (i.e. involve a 'functional' verb as the higher predicate), the MTC is fully supported, since the only option is for the 'controlling' subject to be generated as the subject of the embedded predicate and moved from there to the higher subject position in which it surfaces. The reason is that for the large majority of restructuring predicates, it holds that they are devoid of independent thematic properties (i.e. they do not assign an external theta-role). This is shown by the fact that these predicates (functional heads in Cinque's analysis) fail to impose selectional requirements on the subject of their clause (Cinque 2004):

(1)   a.   La casa gli doveva piacere.
            'the house had to appeal to him'.
       b.   La casa non gli poteva piacere.
            'the house could not appeal to him'.
       c.   La casa gli smise di piacere.
            'the house stopped appealing to him'.
       d.   La casa gli stava dando molti dispiaceri.
            'the house was giving him a lot of trubles'.
       e.   La casa gli finì per piacere.
            'the house ended up being appealing to him'.

However, there is a major empirical problem. A subset of restructuring verbs

---

[5] See Rizzi (1978, 1982).

does actually impose selectional requirements on the subject, as is shown in (2) (Cinque 2004):

(2)   a.   *La casa gli voleva appartenere.
              'the house wanted to belong to him'.
        b.   *La casa non gli osava piacere.
              'the house did not dare to appeal to him'.
        c.   *La casa non gli sapeva piacere.
              'the house didn't know how to appeal to him'.
        d.   *La casa gli provò a piacere.
              'the house tried to appeal to him'.

This seems to indicate that restructuring is compatible, after all, with a situation in which the higher 'functional' verb does actually assign an external thematic role. On these grounds, if one still wants to endorse the MTC, she is forced to adopt Hornstein's non-conservative version of the theory, according to which a single argument can be associated with more than one theta-role and theta-roles are configurationally assigned both through External and Internal Merge, with the result that moving the subject from the lower to the higher position is tantamount to endowing it with two distinct theta-roles.

Cinque takes a different route. He observes that the data in (2) are at odds with other data (based on *ne*-extraction and the properties of impersonal (-passive) *si* constructions), which show that even the verbs in (2) (and not only those in (1)) do not take any external arguments. In other words, Cinque still pursues the idea that Raising provides a uniform analysis for all the control structures that admit restructuring. He suggests that if this view could be maintained, there would be important conceptual advantages. In particular, he adopts Wurmbrand's (2002) analysis, according to which the divide between EC-predicates and PC-predicates is determined by restructuring: exhaustive control (EC) is uniformly compulsory for restructuring control predicates and partial control (PC) uniformly available for non-restructuring control predicates. If this is correct, and restructuring uniformly entails raising, the explanation would consist in the fact that exhaustive control is a mechanical side-effect of raising (in fact, there would be a unique argument with a unique theta-role in these structures), whereas partial control would follow as the default semantics assigned to non-restructuring control predicates, involving PRO as the subject of the embedded predicate.

There is, however, an important residual problem to be solved: how to account for the data in (2), which suggest that certain restructuring predicates do in fact assign an external theta-role? Let us dub the contrast between (2) and the data suggesting that the subject of restructuring predicates is uniformly a 'derived' subject as *Cinque's paradox*. In a nutshell, it can be formulated as follows: How can we enforce a raising analysis for restructuring predicates while maintaining that some of them still express selectional requirements on their subject? Cinque's solution to the paradox (inspired by Zubizarreta's (1982) notion of adjunct theta-role) is not very principled and consists in "taking their selectional requirements to be a consequence of their semantics. If verbs like 'want'... must be predicated of a sentient being, the ungrammaticality of (55) [our (2)]... follows without having to assume that they take an external argument of their own" (Cinque 2004: 15). As is evident, what remains to be explained is why semantic selection requirements that are normally identified as theta-roles are not formally expressed as theta-roles, to be syntactically discharged. In a sense, this solution is tantamount to stipulating that there are in fact two theta-roles assigned to the same argument. And if this is the case, we lose the main motivation for the proposed conservative stand towards 'generalized' MTC, in which arguments may receive more than one theta-role.

In the next section, we will propose a principled solution to *Cinque's paradox*, based on the idea that the semantics of the four predicates in (2) enforces a lexical process of thematic overwriting whereby the higher theta-role overwrites the lower theta-role[6]. We propose that it is the availability of thematic overwriting that allows the syntactic realization of these predicates as 'functional' heads rather than as regular lexical verbs: as proposed by Cinque, they directly lexicalize a functional position within a given functional hierarchy, and cannot head a VP of their own. However, it will turn out that the semantic requirements for thematic overwriting are also satisfied by some non-restructuring control predicates, and that these predicates naturally admit Partial Control. It follows that Cinque's generalization (based on Wurmbrand 2002), according to which PC is a function of the semantics of PRO, is not empirically sustained, since thematic overwriting is incompatible with the syntactic realization of the embedded theta-role as PRO.

An explicit and principled solution to Cinque's paradox opens a completely different scenario on the nature of (Partial) Control. In the next section, we introduce the basic ingredients of this scenario.

---

[6]This is based on the notion of an active lexicon and complex lexical operations especially envisaged (and partially developed) in Tanya Reinhart's work (Reinhart 2002). More particularly, the operation of theta-overwriting we propose in this contribution, though conceptually connected to Reinhart's operation of theta-bundling, should be kept distinct from the latter.

3 Immunity to error through misidentification as a trigger for theta-overwriting

The insight we intend to develop in this section is basically the following: the four verbs that resist a direct raising analysis in Cinque's scenario (*sapere* 'know, *volere* 'want', *osare* 'dare', *tentare* 'try') can be made compatible with a unique argument analysis in terms of A-movement (i.e. Raising), provided one accepts the idea that the theta-role assigned by the higher predicate[7] *overwrites* the role assigned by the subordinate predicate. In this way, only one theta-role is syntactically active, which can be realized, as far as these predicates are concerned, in the lower subject position. As a first approximation, thematic overwriting can be defined as in (3):

(3) $\lambda x \lambda y [\text{Exp}_{V1}(x) \ldots \text{Ag}_{V2}(y)] \rightarrow \lambda x [\text{Exp}_{V1+V2}(x)]$

Roughly, (3) is intended to express the insight that in the cases at hand the lower theta-role is deleted for the aims of the *syntactic* computation (correctly deriving the EXPERIENTIAL reading associated to the IEM-effects), while still remaining available, as we will see, for the systems of interpretation. Thematic overwriting arguably provides a conceptual solution to Cinque's paradox. However, as it stands now, the proposed analysis simply makes explicit, by means of the specific lexical operation in (3), what was left 'implicit' in Cinque's analysis, through the opaque notion of 'adjunct theta-role' or, alternatively, a not less opaque notion of theta-roles that need not be realized in syntax. The situation is different if we can show that there exists an independent trigger for thematic overwriting with restructuring control predicates. If there are independent reasons for (3) to apply, the raising analysis of these control structures is no longer based on the stipulation that one of the relevant theta-roles is not assigned in syntax, but can be seen as the result of the application of a well-motivated set of interface conditions (i.e. the condition – to be established – that makes overwriting possible on general UG grounds, and the condition – already established – that makes overwriting an effective solution to *Cinque's paradox*). So, the question is now: Which general interface principles make thematic overwriting possible in the case of restructuring control structures?

We believe that the answer to this question lies in the semantics of the relevant predicates, more exactly in the fact that they license a variety of *de se*

---

[7] The higher predicate corresponds with the verb inserted—under Cinque's analysis—into a dedicated position within the clausal functional hierarchy, since its meaning fully matches the meaning universally associated to that functional position.

reading that is commonly referred to, in the philosophical literature, as 'immunity to error through misidentification' (IEM). The phenomenon has originally been described in (Wittgenstein 1958) with respect to the peculiar sort of first-personal interpretation assigned to the first-person pronoun ('subject-I' in Wittgenstein's terminology) in sentences like those in (4) below:

(4)   a.   I am in pain.
       b.   I see a canary in the room in front of me.

As noticed by Wittgenstein, it does not make sense, in these cases, to inquire about the identity of the subject of the described experience by asking, for instance: "Is it really you who is in pain"? or: "Is it really you who is seeing a canary?". The reason for this is, intuitively, that (4) holds as a direct report on a phenomenal experience which is *immediately given*, that is, not based on acts of reflection or on external perceptual data[8]. In this sense, the IEM-reading of (4) is only one of the possible interpretations. A non-IEM-reading of (4) arises, for instance, in a context in which I, being presented with two pictures featuring a man whom I come to identify with myself, might describe their content by uttering the sentence: "Here, I am in pain" (if the man in the picture displays for instance overt signs of physical distress) or the sentence: "Here, I see a canary in the next room" (if the man in the picture is portrayed as glazing at a canary in front of him). In the literature, the IEM-reading has been referred to as a kind of 'implicit *de se*'[9], characterized by the absence of an explicit process of becoming aware of the identity between the object(s) of thinking and the subject of thinking, or in terms of the basis relativity of the judgment involved, as is made clear by the following quote:

(5)   There are different bases on which I might judge 'I hear trumpets'. For example, I might base that judgment on an auditory perception I am having. If I do, the judgment seems to be fp-immune [*i.e. immune to error through misidentification relative to the first person*]. But what if I had made the same judgment on the basis of an inference from the following judgments: 'The person in the third row hears trumpets' and 'I am the person in the third row' (we can imagine that I have come to know both of these premises through testimony)? At least when made on this kind of inferential basis my judgment 'I hear trumpets' does not seem

---
[8] See Shoemaker (1968).
[9] See Recanati (2007).

fp-immune. If the first premise of the inference had been true, but the second premise had been false, the error I made would have been an error through misidentification. (Morgan 2012: 106)

Higginbotham (2003) argues that the subject of control complements is immune to error through misidentification with predicates such as *remember*, *imagine*, and *want*, in the same way as the first person pronoun when used as a subject, in Wittgenstein's sense.[10] Consider the sentences in (6):

(6) a. I remember saying that John should finish his thesis by June.
　　b. I imagine flying.
　　c. I want to solve the problem.

If we apply the diagnostics proposed above for the detection of IEM-effects, it clearly makes no sense, in all three cases above, to inquire about the identity of the subject of experience. More particularly, it is pointless to inquire whether it's truly me the person of whom I remember that he said that John should finish his thesis, whether it is truly me the person of whom I am imagining that he is flying, or whether it is truly me the person who – in my intention – should solve the problem. As above, the reason is that the sentences in (6) uniformly count as reports of phenomenal experiences in which the object of remembering, imagining and wanting is immediately given as the subject of the experience itself, without any reflective or perceptual act of identification or any basis-relative judgment, in Morgan's sense.

Let us consider now in some detail the way in which IEM-effects extend from simple sentences to complex control sentences, contributing to explain not only the contrast detected between the two sentences in (7) but also, crucially, the contrast arising between the two sentences in (8):

(7) a. I sent the letter.
　　b. I am in pain.

(8) a. I want that Mary solves the problem.
　　b. I want to solve the problem.

The speech act consisting in assertorically uttering (7a) clearly commits me to identify the person who sent the letter as myself, but – under the most common circumstances of interpretation – there is no sense according to which the

---

[10] See Delfitto and Fiorin (2014), Fiorin and Delfitto (2015), Delfitto, Reboul and Fiorin (to appear).

content of my assertion is 'protected' from possible errors of misidentification that might have occurred while I was constructing the empirical basis (in Morgan's sense) of my judgment. In fact, if one collects some evidence that the sender was not me but, say, my wife (and he gets convinced that I might have confused memories about the relevant facts) he might legitimately ask me: "Is it really you who sent the letter?". This procedure would be pointless in the case of (7b): when uttering this sentence, I may be mistaken about a lot of things (possibly including the fact that I am miscategorizing the described phenomenal experience as 'pain'), but I cannot possibly be mistaken about the identity of the subject of this experience, for the very reason that there is no question of identity: the subject of the experience is immediately given to me as a *minimal self*[11], beyond any act of reflection or external perception.[12] In fact, there is a conceivable (though admittedly extreme) scenario in which this *self* reduces—for me while uttering the sentence—to the unique property I describe in uttering the sentence (i.e. the property corresponding to the *minimal* experience of feeling pain).

Now notice that this reasoning naturally extends to the pair in (8). When truthfully uttering (8a), I certainly want Mary to solve the problem but once again, it is not impossible to figure out circumstances in which my judgment might have incurred into an error through misidentification. This can be clarified under the assumption that proper names are endowed with a primary and a secondary intension, as in *two-dimensional* approaches. Suppose that the problem that should be solved is a hard mathematical problem and that what I really want while uttering (8a) is that the problem be assigned to the best mathematician in my research group. It simply turns out that I identified Mary as "the best mathematician", based, say, on some clues I had. However, I am completely mistaken about that, since Mary's mathematical skills are in fact quite limited. Since my will—in the given circumstances—could be correctly described by replacing the proper name in (8a) with the corresponding intension "the best mathematician in my group", it follows that it is perfectly legitimate for someone to inquire into the content of my will by asking the question: "Is it really Mary the person who—in your intention—should solve the problem?". In fact, as we have seen, I might finally concede, under a better assessment of the facts, that the person who I want to solve the problem is not Mary but, say, Anne. Conversely, it is actually pointless for you to ask me, after I uttered (8b), whether it is really me the person who—in my intention—should solve the problem, at least under the IEM-reading of (8b). This is the

---

[11]For the notion of *minimal self* see especially Gallagher (2000).

[12]For an in-depth discussion, see Delfitto, Reboul and Fiorin (to appear).

reading of (8b) according to which my uttering the sentence simply commits me to the truth of the subjective experience that I am reporting by means of (8b), according to which my willingness to solve the problem is given to me without me having to descriptively identify the experiencer of this willingness, through some explicit act of reflection or external perception, that is, through some 'basis-relative' judgment.

If this analysis is essentially correct, in the case of *want*—as well as in the case of *remember* and *imagine*—the control structures to which this verb gives rise amount to the description of a single phenomenal experience. For instance, in the case of (8b), the semantics expressed does not require assigning a distinct referential content to both the higher and the lower theta-role, since this sentence, under the IEM-interpretation, reads as the direct report of a single subjective condition of willingness to be the experiencer of an event of solving the problem. This kind of 'experiential' reading emerges clearly when we consider that it applies also in the case of passive complements, as in (9b), to be compared to (9a):

(9)   a.   John wants to kiss Mary.
        b.   John wants to be kissed by Mary.

The willingness of John to be the experiencer of an event of kissing Mary in (9a) translates into the willingness of John to be the experiencer of an event of being kissed by Mary in (9b), quite independently of the different properties of the lower theta-role. More particularly, notice that the theta-role of 'kiss' that gets overwritten in (9b) is the patient theta-role, which is generally characterized as lacking any reference to mental states.[13] On these interpretive grounds, it really makes sense to re-interpret the lexical operation that was described in (3) and restated in (10) as an *asymmetric overwriting operation*, whereby the theta-role associated with the higher predicate (the experiencer) *overwrites* the theta-role expressed by the lower predicate (which is thus cancelled by the higher experiencer):

(10) $\lambda x \lambda y [\text{Exp}_{V_1}(x) \ldots \text{Ag}_{V_2}(y)] \rightarrow \lambda x [\text{Exp}_{V_1+V_2}(x)]$

As we will see in section 6, this analysis in terms of thematic overwriting also offers important advantages for a principled analysis of Partial Control.

Let's take stock and briefly consider where we are. We started by asking whether there is some independent grammatical condition enforcing theta-

---

[13] See Reinhart's (2002) theta-system, where the 'patient' role translates into the feature pair [- mental, - cause].

overwriting as a solution to *Cinque's paradox* and we have been able to find some cases of control in which there is actually no support whatsoever for the syntactic realization of two distinct theta-roles endowed with independent referential content. These structures exhibit strong IEM-effects, whereby the experiencer of a unique phenomenal experience is immediately given as a *minimal self*. We propose that UG favors theta-overwriting in these structures, that is, the application of an overwriting procedure that cancels the lower theta-role for the aims of the syntactic computation, replacing it with the higher theta-role (the experiencer). The intuitive reason for this is that projecting two or more theta-roles into distinct argument slots only makes sense if these theta-roles have any prospects to translate into distinct referential indexes. When control structures are interpreted in terms of reports on phenomenal experiences, to the effect that the referent of the predicates involved is necessarily a *minimal self*, this condition is clearly not satisfied, since no referential index can be assigned except from the one assigned to the minimal experiencer.

If we could show that all verbs that turned out to be problematic for a raising analysis (based on the observation that the higher predicate has thematic properties) give rise to IEM-effects in control structure, we would have a principled motivation for *overwriting* to apply in these structures, reducing the two theta-roles to one and providing an elegant solution to Cinque's paradox. On the one hand, since overwriting provides an effective solution to *Cinque's paradox*, we would end up with the desired analysis of restructuring in terms of raising. On the other hand, we would end up with the discovery that it is the availability, for a certain subclass of control structures, of a special type of *de se* reading, granting immunity to error through misidentification, that guarantees a uniform analysis of restructuring as raising.

We have already shown that IEM-effects are clearly detectable with want/*volere*. More explicitly, we propose now that *volere*, as a restructuring verb in Italian, is directly inserted in a dedicated functional position that exactly matches—as a consequence of the UG format—its semantic content. The external theta-role of *volere* overwrites the subject theta-role of the embedded predicate, under the conditions discussed above. The unique theta-role so obtained can then be realized in the lower subject position, and undergoes raising under standard syntactic assumptions (say, for case reasons). This entails that under an IEM-account, *volere* is virtually indistinguishable, on syntactic grounds, from the restructuring predicates that give rise to control structures where there is no external theta-role to be assigned.

Let us now consider the case of *tentare* (try), by evaluating sentences like (11), as discussed in Grano (2011)—see also Pearson (2013):

(11)   John tried to go to the movies.

Based on the references above, two ingredients may be deemed necessary to provide a convenient semantics for (11): (i) first, the event of going to the movies must have started being realized to some degree; in this sense the semantics of *try* is partially modeled on the semantics of the progressive; (ii) second, the fact that John tried to go to the movies entails that John wanted/intended to go to the movies. As a confirmation of (ii), Grano adduces the felicity of (12):

(12)   John did not try to cross the street; he crossed the street accidentally.

Moreover, (i) and (ii) are strictly intertwined. Consider the case of a person who is physically incapacitated to move (cf. Pearson 2013), in a context in which she is attempting at opening a door. At the moment in which she is still in the process of activating her motor system in order for her arm to reach out for the doorknob (possibly without success), we cannot felicitously utter (13a) if her arm has not moved yet, whereas uttering (13b) sounds perfectly sound:

(13)   a.   Mary is opening the door.
       b.   Mary is trying to open the door.

The point to be made here is that having the intention to open the door, at a stage where this intention has already been 'put into action', that is, it has already been translated into some neurophysiological state associated with the activation of the motor system, already counts as 'trying', though no physical movement has taken place yet. Crucially, the observation that 'trying to open the door' entails 'wanting to open the door' makes control structures with *try* as the higher predicate optimal candidates for a *de se* reading with IEM-effects. And in fact, if I truly utter the sentence "I'm trying to open the door", it seems pointless for you to inquire into the possibility of an error through misidentification, by asking something of the sort: "Are you sure you are correct in identifying yourself with the person who is trying to open the door?". The reasons for this are the same as for the cases already examined: "I'm trying to open the door" counts as a direct report on a phenomenal experience whose subject is immediately given as a *minimal self*: there is no process of 'descriptive' identification of the experiencer based on external perceptual evidence or reflective strategies, hence it makes no sense to inquire about the correctness of the identification procedures applied in the course of this process.

We think that the same type of considerations readily extends to *osare* 'dare' and *sapere* 'know'. Take a sentence like (14a):

(14) a. Oso metterlo per iscritto.
'I dare to write it down'.

There is a reading according to which by uttering (14a), I am truthfully reporting on a unique phenomenal experience that is immediately given to me, consisting in writing down something with the feeling that I am showing, in doing this, a considerable degree of courage or defiance. Under this IEM-reading, there is no sensible question concerning a possible identity mismatch between the person who feels courageous/defiant and the person who is writing down something.

Let us now consider (14b):

(14) b. So risolvere l'equazione.
'I am able (lit. know) to solve the equation'.

Clearly, the meaning of this sentence comes very close to its equivalent with a modal. It readily translates in English as "I can/am able (to) solve the equation". We should thus not be surprised that this control structure is a raising structure, as is the case with modals. Still, we observed above that *sapere* assigns a subject theta-role. So, *sapere* cannot be used when the subject is an inanimate object, as in "*Questa barca sa galleggiare" ('this boat knows to float'), where only the modal is allowed ("Questa barca può galleggiare"). It seems that in this case the presence of a subject theta-role simply reduces to the selectional requirement that a mental state be involved. This entails that in uttering (14b) I am making reference to an inner/mental state that allows his/her bearer to be able to solve the problem. Again, no sensible question can be posed about a possible referential mismatch between the bearer of this mental state and the agent involved in solving the problem.

Let us see which conclusions are warranted. There are some 'functional' verbs that give rise to restructuring and uncontroversially assign a subject theta-role. Since these verbs do not project a VP, how can this theta-role be syntactically accommodated? It turned out that these are cases where control structures are not simply associated with *de se* readings, but are actually associated with a specific subclass of *de se* readings whereby the bearer of a mental state *implicitly* and *immediately* identifies herself with the object involved in this mental state, without perceptual and reflective grounds, and cannot thus be held as mistaken about this identification (IEM-reading). *We have further proposed that UG does not care about the independent realization of two theta-roles when they cannot possibly bear distinct referential indexes.* The consequence is theta-overwriting, which cancels the lower theta-role from the syntactic derivation (while keeping it alive for the systems of interpretation). As a consequence, the only experiencer theta-role of a complex predicate of phenomenal experience can be syntactically realized in the lower subject position and then moved to the higher subject position, as is the case in canonical raising. Moreover, as we will see in section 5, this solution paves the way for a generalized analysis of Control that preserves some advantages of the MTC, while virtually solving the serious difficulties that the MTC encounters.

4 Theta-overwriting and lexical reflexives

Before we outline some of the consequences of the analysis proposed above for a generalized theory of control, we would like to discuss some facts regarding lexical reflexives (in English and especially Dutch) that arguably provide an independent confirmation of one of the main hypotheses put forward in the preceding sections: in contexts where two theta-roles cannot possibly be assigned distinct referential indexes, theta-overwriting applies as the default UG option, and gives rise to IEM-effects when the higher theta-role at stake is an Experiencer. In this section, we show that overwriting also takes place in contexts that do not express propositional attitudes.

First of all, let us consider the interpretation of self-reflexives in (15a-b), for English and Dutch respectively:

(15)   a.   Bill admired himself.
           b.   Bill bewonderde zichzelf.

In (15), we detect the same kind of *de re / de se* ambiguity that arises in the complements of verbs of propositional attitudes (Delfitto and Fiorin 2008). To briefly illustrate this, consider the scenario (inspired by Castaneda's (1968) 'war-veteran' classical setting) in which Bill is watching a man on TV who is bravely rescuing a boy whose life is endangered. Bill admires the man, without realizing that what he is presently seeing is the recorded images of something he himself did years before. On analogy with Castaneda's case of *de re* readings in contexts of propositional attitude, an external observer is allowed to report this situation, in English or Dutch, by making use of (15). The reason for this can be clarified by making use of the notion of 'acquaintance relation' (Kaplan 1989; see also Maier 2010 and the references cited there) as the source of the ambiguity. Though the admired person is certainly Bill (coreference is induced by the use of the reflexive pronoun, under standard assumptions), the *res* he is admiring is always accessed by means of an acquaintance relation (for instance, in the case at stake, 'the man who is bravely rescuing the boy'), in terms that are thus compatible with Bill's unawareness that the *res* is Bill himself. The same considerations hold for the most readily available *de se* reading of (15): Bill admires himself, in the full awareness that the person whom he admires is he himself. In this case, we can assume that the *res* Bill is admiring is accessed by Bill by means of the acquaintance relation 'identical to Bill': under 'identity' as the salient acquaintance relation, a *de se* reading is promptly enforced.

Remember now that Chierchia correctly pointed out[14] that *de se* readings can be grammatically enforced: control structures are generally not amenable to *de re* interpretations. If reflexives give rise to the same sort of ambiguity, we may expect that there are cases of reflexivization where a *de se* reading is grammatically enforced as well. The prediction is borne out. Such a case is provided by a subclass of reflexives in Dutch, exemplified in (16):

(16) a. Jan verbaasde zich.
          'John got surprised'.
      b. Jan bewoog zich.
          'John moved'.

There are no scenario's in which (16a) could be used with a *de re* interpretation, to express, for instance, the reading according to which Jan surprised himself in seeing his own image reflected in a mirror, and without recognizing the image in the mirror as he himself. The only possible interpretation of (16a) is one in which it is used as a report of the fact that Jan got surprised. What is reported is a past phenomenal experience of surprise whereby the cause of the surprise cannot be distinguished from the experiencer: the experience is immediately given to a *self* that is not identified by means of explicit acts of reflection or perception. If I utter a sentence like "Ik verbaas me" ('I am surprised'), it is thus pointless for you to inquire whether I could be mistaken about the identity of the experiencer of the surprise, as an instance of error through misidentification. In fact, it makes no more sense for you to ask me: "Are you really sure that it is you who is surprised?" than it does when you ask: "Are you sure that it's really you who is seeing a canary?" as a reaction to my assertive use of the direct experience report "I see a canary". On these grounds, (16) can be taken to instantiate the subclass of *de se* readings that we have identified as IEM-readings. As a confirmation, let us examine the interpretive relation that a sentence such as (16b) entertains with the transitive and unaccusative variants of 'bewegen' (move). The three relevant sentences

---

[14] See Chierchia (1989) and much subsequent literature.

are given in (17):

(17) a. Jan bewoog de gordijnen. (transitive)
'John moved the curtains'.
b. De gordijnen bewogen. (unaccusative)
'the curtains moved'.
c. De gordijnen bewogen zich. (reflexive)
'the curtains moved-refl'.

In the unaccusative and reflexive variants in (17b-c) the subject is inanimate. The unaccusative variant in (17b) expresses the meaning change traditionally associated with causative alternation phenomena: the external theta-role is suppressed and what is put in the foreground is the change of state that the curtains undergo, i.e. the transition from the state in which they were motionless to the state in which they are moving. The interpretation of (17c) is different: this sentence unavoidably evokes a sort of 'ghost-effect', since by uttering it the speaker somehow entails that the curtains have an inner power/disposition to move themselves[15]. How can we account for this odd interpretive effect? Suppose that lexical reflexives like 'zich bewegen' necessarily involve theta-overwriting, under the version developed above. This would mean that the external theta-role (say, 'Cause') *overwrites* the Patient theta-role: what moves must thus retain the properties of a Cause. The same phenomenon arguably takes place in the case of 'zich verbazen', in sentences such as (16a). Here the two theta-roles involved, under standard assumptions, are Cause and Experiencer, as made evident by the transitive variant "Zijn beeld in de spiegel verbaasde Jan" ('his image in the mirror surprised John'). Under theta-overwriting, one of the two theta-roles *overwrites* the other, yielding the interpretation of (16a) according to which the Cause is indistinguishable from the Experiencer himself. What is reported is a phenomenal experience whereby the experiencer of the surprise is the cause of her own surprise. What explains the interpretive difference between (17b) and (17c) is thus the fact that in the reflexive variant the Cause role is not eliminated (as in the unaccusative variant) but rather overwrites the Patient role, yielding the odd reading according to which the curtains do not simply undergo movement but must somehow be conceived of as causing their own movement. The conclusion we'd like to draw is that reflexive structures such as (16a) instantiate an independent case of metaphysical indistinguishability of the two theta-roles involved, characterized by the impossibility of associating these theta-roles with

---

[15] See Reuland and Marelj (2013) and the references cited there.

distinct referential indexes, as in the cases of the control structures giving rise to an IEM-reading.

However, one should also acknowledge that the case in (16b) is slightly different, since it is not clear at all that (16b) can be read as the report of a phenomenal experience. This is indirectly confirmed by our discussion about (17c): the 'ghost-effect' detected here simply consists in viewing the curtains as endowed with a causal potential for movement, and not in viewing them as endowed with a potential for counting as bearers of phenomenal experiences. So, the question is: Since (16b) is not a case of IEM, what exactly enforces theta-overwriting in cases such as (16b)?

An answer to this question can be found by examining other canonical cases of lexical reflexives in English and Dutch, like those in (18):

(18) a. John washed/shaved.
      b. Jan waste/schoor zich.

Here, there is a clear interpretive difference with respect to the reflexive counterparts of (18) that involve self-reflexives, as in (19) below:

(19) a. John washed/shaved himself.
      b. Jan waste/schoor zichzelf.

The difference does not consist in the fact that the lexical reflexives in (18) are necessarily read *de se*, whereas the sentences in (19) are ambiguous between a *de re* and a *de se* interpretation. There is certainly a strong bias to interpret (18) as a report on an event in which John consciously shaved himself, i.e. he shaved while being completely aware that the person being shaved was he himself. In spite of this, it is not difficult to figure out situations in which (18) can be read in ways that are incompatible with a strict *de se* reading. For instance, suppose that John is found in the bathroom in a sleep-walking condition while engaging in a series of actions that we would qualify as shaving. In this scenario, we are allowed to report the situation, both in English and Dutch, by using sentences containing a lexical reflexive. We might say, for instance: "I entered the bathroom and I saw John shaving", whereas it is quite likely, in the sleep-walking scenario, that John was not acting consciously, i.e. he was probably not aware, among other things, that the person being shaved was he himself.

The correct generalization seems thus to be that in the structures involving lexical reflexives there are strong reasons to identify one of the two arguments of the predicate as referentially non-distinguishable from the other. In the case of 'zich verbazen', there are no reasons to distinguish between the 'experiencer' argument and the 'cause' argument, since what we are reporting is an immediately given phenomenal experience. In the case of 'zich bewegen', there are no reasons to distinguish between the 'cause' argument and the 'patient' argument: if we utter the sentence "Jan bewoog zich', what we are reporting is a series of automatically coordinated motor control instructions that produces the effect that the object that referentially counts as the locus of these instructions is put into movement. Roughly the same effect is found in the case of 'zich wassen/scharen': here, it is the Patient and the Agent argument that are merged together. What is described is a state of affairs in which the Agent automatically performs a well-defined series of actions (including internal motor control instructions) automatically affecting some of his body parts (similar considerations hold for 'zich ontkleden' (to get undressed), etc.). In this case, it is thus not required that the Agent be aware that she is non-distinct from the Patient, it is only required that the agent be involved in an automatically developing course of actions whereby there is no sensible distinction to be made between the Agent and the Patient, in the sense that the individual being washed/shaved cannot possibly be different from the individual who performs the given course of actions.

A strong confirmation that this is essentially the correct analysis is provided by the literature on lexical reflexives,[16] which emphasizes that so-called *proxy-readings* are completely excluded with lexical reflexives, whilst they are allowed for the variants involving self-reflexives. In a wax-museum scenario, in which John is moving, washing, shaving his wax-counterpart, we cannot report this situation by means of the sentence "John beweegt/wast/scheert zich", whilst it is acceptable to describe it by means of the sentence "Jan beweegt/wast/scheert zichzelf". Under the analysis proposed above, this immediately follows. Proxy-readings do not satisfy the crucial requirement for theta-overwriting, that is, the condition according to which overwriting is enforced by UG to take place between two given theta-roles when these two theta-roles cannot possibly be assigned *distinct* referential indexes. The statue of John goes proxy for John and it is not exactly John. In fact, the sort of movements required from John when he washes his wax-counterpart (including the internal motor control instructions) are quite different from the movements (again, crucially including the internal motor control instructions) that result in 'John washing'.

On these grounds, we conclude that the data on lexical reflexives corroborate the hypothesis according to which theta-overwriting is automatically activated, on UG grounds, whenever two theta-roles are referentially indistin-

---

[16] See Reuland (2011) and the references cited therein.

guishable. This can result in *de se* readings displaying IEM-effects (when an Experiencer role is involved) or in readings where metaphysical indistinguishability provides the grounds for an interesting extension of canonical IEM effects to non-experiential contexts, as is the case for "zich bewegen/wassen/scheren". If I truthfully utter the sentence "I shaved", it cannot be the case that I was mistaken in identifying the shavee with myself. The reason is that if the shavee is not me but another person, it necessarily follows that I miscategorized the whole shaving event (since, as discussed above, the event of 'shaving' is distinct from the event of 'shaving someone', including the event of 'shaving himself', that readily allows a proxy-reading). In all cases, theta-overwriting is dictated by *metaphysical necessity*, i. e. by the impossibility that the theta-roles involved be distinguished referentially. However, *de se* effects manifest themselves only when an Experiencer theta-role is involved, giving rise to reports on phenomenal experiences. The conclusions reached in the preceding section on IEM-effects as a trigger for overwriting still stand, but they are now part of a scenario in which IEM-effects and *de se* readings are not necessarily associated.

5 IEM-effects, PRO and Partial Control

In section 2, we have seen that Cinque (2004) identified the class of restructuring verbs (including those that assign a subject theta-role) with the class of raising predicates. Since Raising entails the presence of a unique subject argument, exhaustive control effects (EC) are predicted for this class of control structures. The other control structures involve the presence of PRO as the subject of the embedded clause. Partial Control (i.e. the possibility that the referent of PRO properly includes the referent of its controller) follows then naturally as a stipulated ingredient of the semantics of PRO.

Undoubtedly, this would count, if correct, as a quite elegant analysis. Unfortunately, there are insurmountable empirical difficulties with it. First, certain restructuring verbs that assign an external theta-role admit Partial Control. This is the case with *volere* (want). Desideratives like 'want' are well-known as PC-verbs in English.[17] Here are some examples:

(20)   a.   John wants to meet at 9 am.
       b.   John wants to go on holiday all together.
       c.   John wants to work at the problem as a team.

---

[17] See Pearson (2013) and the references cited therein.

d. John (an architect) wants to build the new town without endangering the environment.

Some of these examples translate perhaps into Italian (judgments are not uniform among speakers), in sentences where *volere* is used:

(21) a. Gianni vuole riunirsi alle 9.
b. Gianni vuole andare in vacanza tutti assieme.
c. Gianni vuole lavorare al problema come team.
d. Gianni (un architetto) vuole costruire la città senza danneggiare l'ambiente.

There is clear evidence – we believe – that (21a) is a 'fake' example of PC, and rather involves an 'empty comitative analysis'[18]. For instance, English (22a) cannot translate into Italian as (22b), where the collective predicate 'riunirsi' is used, but as (22c), where the predicate has a singular 'comitative' meaning that is in fact compatible with the overt realization of the comitative argument:

(22) a. I want to meet at 9 am.
b. *Voglio riunirsi alle 9.
'I want to meet-PL at 9'.
c. Voglio riunirmi alle 9 (con gli altri membri del gruppo).
'I want to meet-SG at 9 (with the other members of the group)'.

Independently of the status of (21a), we will assume, for the purposes of this contribution, that the examples in (21b-d) are sufficient to show that *volere* can give rise (perhaps only marginally) to PC effects. Since *volere* is a restructuring verb, this fact is incompatible with Cinque's hypothesis.

Second, and less uncontroversially, there are non-restructuring verbs that do not admit PC. This is the case for 'propositional' verbs such as 'believe', 'claim' and 'pretend' in English.[19] These data readily translate into Italian, as shown by (23):

(23) a. *Gianni crede di essere andati in vacanza tutti assieme.
'John believes to have gone-PL on holiday all together'.
b. *Gianni sostiene di essere andati in vacanza tutti assieme.
'John claims to have gone on holiday all together'.
c. *Gianni pretende di essere andati in vacanza tutti assieme.
'John pretends to have gone on holiday all together'.

---

[18] For this account of PC-phenomena, see Boeckx, Hornstein and Nunes (2010).
[19] See Pearson (2013).

Since this subset of propositional verbs does not give rise to restructuring effects – hence a raising analysis is impossible for them – this behavior is incompatible with Cinque's hypothesis: these structures are predicted to involve the presence of PRO as the subject of the embedded clause, and the stipulated semantics of PRO should make PC readings possible.

Third, the discussion in the preceding section strongly suggests that the possible separation line between structures not involving PRO (thus incompatible with PC) and structures involving PRO (by hypothesis, compatible with PC) should not be drawn between restructuring verbs and non-restructuring verbs, but rather between the class of structures characterized by raising and theta-overwriting on one side, and the class of verbs that do not permit theta-overwriting on the other side. More particularly, the unitary analysis of restructuring verbs as involving raising was shown to be a side-effect of the IEM-reading triggered by the restructuring predicates that assign an external theta-role. For these predicates raising is parasitic – so to speak – on theta-overwriting, which represents the grammatical encoding of IEM-readings. If this is correct, a clear-cut prediction is made: we expect that control structures where *de se* gives rise to IEM-effects should be incompatible with PC. The reason is that an IEM-reading is based on the application of thematic overwriting, and overwriting typically reduces two distinct theta-role to a unique Experiencer role, triggering raising instead of the realization of a referentially distinct PRO argument in the lower subject position. At first sight, this prediction does not seem to radically change the nature of Cinque's conjecture: the (slightly revised) prediction is that PC should be incompatible not only with all monothematic raising predicates (typically, aspectual and modal verbs, as Cinque originally suggested) but also with all verbs that give rise to IEM-readings, crucially including 'implicatives' such as 'dare' and 'try' (cf. the analysis in section 3). The rest of control structures is predicted to involve PRO, hence to give rise to PC effects. Quite remarkably, moreover, the empirical refinement under discussion seems to provide an important conceptual advantage. It might provide a principled reason why PRO triggers PC-effects (consider that this remained a stipulation in Cinque's original account). In a nutshell, the reason is the following. As we have seen in section 4, theta-overwriting is incompatible with proxy-readings. This directly follows from the nature of overwriting, which is triggered by the referential indistinguishability of the theta-roles involved. Conversely, it is well-known that pronominal elements readily lend themselves to proxy-readings: this is the case for personal pronouns, crucially also under a bound-variable reading, and for the self-anaphors discussed in section 4.[20]

---

[20] See Reuland and Winter (2009) for a formal analysis of proxy-readings in terms of Skolem

Under this premises, a non-stipulative account for the reason why PRO triggers PC is immediately available: it suffices to regard the *extension phenomenon* by means of which the referent of PRO comes to include the referent of its controller as a straightforward instance of proxy-reading for PRO. Since PRO is a pronominal/anaphoric element after all, this is exactly what should be expected on general grounds (however, see the next section for some important qualifications).

Unfortunately, in spite of its apparent plausibility, this empirical/conceptual refinement of Cinque's conjecture, inspired by our analysis of the role of IEM-effects for the syntactic analysis of control, is empirically untenable. The reason is that if we draw the division line as proposed above, too many predicates would fall on the wrong side of the line. For instance, there is no doubt that propositional predicates like 'remember' and 'imagine' give rise to strong IEM-effects.[21] Unfortunately, it is also uncontroversial that these propositional predicates are pretty much compatible with PC, as shown by (24):

(24)   a.   I imagine working at the problem all together.
            b.   I remember going on holiday all together.

Moreover, not only propositional predicates such as 'remember' and 'imagine' are problematic (PC effects without PRO), but also propositional predicates like 'believe', 'claim' and 'pretend' (EC with PRO), since – as noticed above – the latter do not seem to give rise to IEM-readings, hence theta-overwriting is not expected to apply.

On these grounds, two conclusions seem inescapable: (i) the semantics of PC cannot be discharged on the semantics of PRO (*pace* Cinque); (ii) IEM-effects extend quite beyond the class of restructuring predicates. An important consequence of (ii) is that (under the hypothesis that IEM-readings involve theta-overwriting) thematic overwriting extends beyond the class of restructuring predicates.

In this respect, there are essentially two questions to be addressed. First, what is the syntax of non-restructuring overwriting predicates? Second, what is the relationship between overwriting and PC? In the remainder of this section, we will address the first question, while the second issue will be discussed in the next section, where we will propose some new original insight about the nature of Partial Control.

As for the first issue, it seems to us that the default hypothesis is that if a non-restructuring predicate involves overwriting, the unique argument of this

---

functions.

[21] See Higginbotham (2003) and the discussion in Delfitto and Fiorin (2014).

predicate must be realized in the subject position of the higher predicate. Here is why. The higher verb in a non-restructuring structure is by definition a lexical verb that cannot be inserted into a matching head position within a rigidly defined hierarchy of functional categories. It projects thus a full VP, hence a thematic subject position where the external theta-role can be syntactically realized. Once overwriting has taken place, reducing the number of arguments that must be syntactically realized in the control structure to one, two options are available: (i) the unique theta-role is syntactically realized in the lower subject position and then moved to a higher subject position; (ii) the unique theta-role is directly realized in the higher subject position, whereas the lower subject position is simply not projected in syntax. There are strong reasons to prefer option (ii) to option (i). Option (ii) is conceptually straightforward: if two theta-roles are reduced to one, the syntactic space is reduced in the most economical way. Conversely, option (i) has to face two major problems. First, it violates elementary conditions on the economy of derivation, since we would have two instances of Merge (one operation of External Merge and one operation of Internal Merge) instead of one (the single operation of External Merge consisting in generating the argument in the higher subject position). Second, the operation of Internal Merge would consist in displacing the subject either directly into a thematic A-position (an option that should be avoided within the 'conservative' version of the MTC inspired by Cinque) or at least by crossing a thematic position (a minimality violation). These complications (independently of whether they can be overcome technically) simply do not arise under option (ii).

Under the analysis proposed above, non-restructuring overwriting predicates such as 'remember' and 'imagine' do not involve raising and do not involve PRO. At the same time, we suggested that the primary motivation for theta-overwriting consisted in providing a principled way out from Cinque's paradox in the case of restructuring control structures. The problem was how we can enforce a raising analysis for restructuring control structures if we need to discharge two independent theta-roles into syntax. The solution was that there aren't in fact two independent roles, since restructuring control structures are arguably interpreted as complex experience predicates involving a unique experiencer. On these grounds, the evidence just reviewed suggests that overwriting might have generalized as a core grammatical device to encode IEM-readings, quite independently of the functional or lexical status of the higher predicate in control configurations. From this perspective, a further possibility is that overwriting further generalizes to the predicates that do not trigger IEM-effects, as a third step in diachronic development. As an example in Italian, consider (25), which involves control but clearly entails the presence of an error through misidentification:

(25)   Gianni pensa/crede di aver spedito la lettera, ma si sbaglia (a spedirla è stata sua moglie).
'John thinks/believes that he sent the letter, but he is wrong (it was his wife who sent it)'.

There is a clear sense that the third step should be the more problematic, since it entails overcoming the original motivation for theta-overwriting, that is, encoding the referential indistinguishability between two theta-roles. And in effect, there is evidence that it *is* more problematic, since parametric variation among languages is typically found at the level of the predicates that do not give rise to IEM-readings. 'Believe' and 'think' represent cases in point—they are incompatible with control in English but not in Italian:

(25)   a.   Gianni pensa/crede di essere intelligente.
        b.   *John thinks/believes to be intelligent.

Clearly, 'credere' is a predicate that typically gives rise, in Italian control structures, to *de se* readings devoid of IEM-effects, as shown by (26) below:

(26)   Credevo di essere stato a Praga nel 1989, ma poi mi sono convinto che ci è stata solo mia moglie.
'I believed that I had been in Prague in 1989, but then I realized that only my wife had been there'.

Suppose that this approach is essentially correct. It would entail that if a language allows control to apply beyond the class of monothematic raising predicates and the class of IEM-predicates, it does that by extending the domain of theta-overwriting beyond the domain of metaphysical indistinguishability between referential positions (i.e. beyond the core domain of IEM-effects). A reasonable alternative would be of course that extended control structures (i.e. control structure that are neither restructuring control structures nor IEM control structures) involve PRO as the subject of the embedded clause (i.e. there is no overwriting in these cases). Here, we leave this issue undecided. However, we should notice that from the present perspective, the only principled reason to introduce PRO would be linked to the possibility of drawing the line between PRO-structures and non-PRO-structures in such a way that the PRO structures are those that allow PC, with the result that the semantics of PC might be seen as a function of the semantics of PRO, with clear conceptual advantages. But we have seen, in the course of this section, that there is no empirically viable way to draw this line.[22]

In the next section, we will address the second problem formulated above, that is, the relationship between theta-overwriting and Partial Control. Facing this problem will allow us to propose some original new insights on the nature of Partial Control and of Control as a whole.

6 The semantics of theta-overwriting and pragmatic enrichment

The issue to be addressed concerns the relationship between the grammatical operation of theta-overwriting and PC. Given the discussion in the preceding sections, there is an important conclusion to be drawn and that should represent the starting point of any analysis of the problem at stake. This is the fact that all classes of predicates that give rise to PC-effects (i.e. propositional, factive and desiderative predicates) contain predicates that manifest IEM-effects. Just to exemplify, 'remember' is a factive predicate, 'imagine' is a propositional predicate and 'want' is a desiderative predicate. Given this observation, the original issue translates into the following: How is it possible that referential indistinguishability of theta-roles (that is, what underlies IEM-readings) be compatible with situations where the referent of the lower theta-role includes the referent of the higher theta-role (that is, with the semantics of PC)? Moreover, there is an important corollary to be emphasized: we not only want to understand how comes that IEM-effects are compatible with PC-effects, we also want to understand why PC manifests itself in control structures independently of the presence of IEM-effects. Namely consider that many desiderative, factive, propositional and interrogative predicates gives rise to PC in contexts where there are no IEM-effects.

A line of analysis that would be compatible with the approach to Control developed here consists in discharging the burden of the explanation on the semantics of the control predicates. A solution of this kind is proposed in (Pearson 2013), where PC emerges as the result of an 'extension effect' within the quantificational structure of control predicates, in the context of a 'property analysis' of the complements of control verbs inspired by Lewis' and Chierchia's analysis of *de se*. The general idea[23] is that the complements of attitude

---

[22] We do not discuss in this paper the independent syntactic motivation that has been proposed in favor of a PRO-like empty category in control structures.

[23] See also Abush (1997).

predicates (on analogy with what is assumed for root clauses) have abstractors over worlds, times and individuals in their left-periphery. This means that these predicates are treated as quantifiers over world/time/individual triples. Exemplifying with the verb 'claim', we get something along the lines of (27):

(27) $[\![\text{claim}]\!]^{c,g} = \lambda P_{\langle e,\langle i,\langle s,t\rangle\rangle\rangle}\lambda x_e \lambda t_i \lambda w_s \forall \langle w',t',y\rangle \in \textbf{claim}_{x,w,t} \rightarrow P(y)(t')(w')$
Where $\textbf{claim}_{x,w,t} = \{\langle w',t',y\rangle :$ it is compatible with what x claims in w at t for x to be y in w' and for t to be $t'\}$

Informally, (27) reads as follows: "*Claim* is the set of properties that hold of the individual y in all the worlds w and times t such that it is compatible with what is claimed by the individual x in the world of evaluation for x to be y, for w to be the world of evaluation and for t to be the time of evaluation". On these premises, Pearson proposes that the reason why some canonical attitude predicates (like 'claim', 'believe' and 'pretend') are incompatible with PC is that these predicates are used to report attitudes about the here and now. These 'simultaneous' predicates contrast with future-oriented predicates (like 'decide', 'want', 'intend', 'hope') and past-oriented predicates (like 'remember' and 'regret'). As originally noticed in (Landau 2000) (and successive work by the same author), it is these future- and past-oriented predicates that typically give rise to PC. This contrast is empirically detectable by investigating the pattern of temporal modification through overt temporal adverbials exhibited by these verbs, as shown in (28) below:

(28) a. *Yesterday/today, John claimed to go to the movies tomorrow.
 (simultaneous)
 b. Yesterday/today, John wanted/hoped to go the movies tomorrow.
 (future-oriented)
 c. Today, John remembers/regrets going to the movies yesterday.
 (past-oriented)

The crucial insight is that 'past-oriented' predicates contain an abstract aspectual operator that shifts the time of evaluation of the embedded clause to the past with respect to the time of evaluation associated with the main predicate. In fact, a sentence like "John remembers going to the movies yesterday" is interpreted as equivalent to the sentence "John remembers having gone to the movies yesterday". Analogously, 'future-oriented' predicates shift the time of evaluation of the embedded clause to the future with respect to the time of evaluation introduced by the main predicate. Finally, it is proposed that factive predicates contain an inherent progressive operator (shifting the time of the embedded clause to a time t' that includes the time t introduced by the main predicate): a sentence like "I am glad to write this paper" is actually read as "I am glad to be writing this paper". Consider now that the reason why a sentence like "*I believe/claim to write this paper" is unacceptable might consist in the fact that 'simultaneous' predicates such as 'claim' and 'believe' do not contain an inherent progressive operator, to the effect that the sentence above cannot be read as "I believe/claim to be writing this paper", unless a progressive operator is overtly introduced in the complement clause, shifting its time of evaluation, as required.

In a nutshell, the conclusion is the following: there are predicates of propositional attitude that manifest a *temporal extension effect*. As seen in (27), these predicates introduce a form of quantification on triplets $\langle w, t, i \rangle$ of worlds, times and individuals. The temporal extension effect consists in the replacement, within this triple, of the variable t introduced by the main predicate with a variable t' that includes, precedes or follows t. On the other hand, there are predicates of propositional attitudes that do not admit any temporal extension effect. The claim is that it is these predicates that are incompatible with PC[24]. The reason—it is submitted—is that, since quantification is on times *and* individuals (cf. (27)), temporal extension goes hand in hand with individual extension. From this perspective, PC is nothing else than the introduction of a containment relation between i' (the individual variable associated with the complement clause) and i (the individual variable introduced by the main predicate). PC is thus correctly predicted to arise only with the verbs that allow temporal extension, since these are also the verbs that allow individual extension. Pearson (2013) further argues that EC-predicates (implicatives, aspectuals and modals) do not allow temporal extension, as expected.

Pearson's analysis is at first sight fully compatible with the analysis of control proposed in the present contribution. More particularly, Pearson's analysis, from the present perspective, has two main consequences: (i) generally speaking, there is no need for PRO in control structure (in full agreement with the conclusions reached at the end of the preceding section); and (ii) PC is a phenomenon that is entirely independent of theta-overwriting and IEM-effects.

In principle, we might subscribe to both conclusions. However, we believe that (ii) is actually *not* correct. We propose that there is in fact an important conceptual link between theta-overwriting and PC. Elucidating this link permits to solve a residual conceptual difficulty implicit in Pearson's account. Let us see why this is the case.

As it stands, Pearson's approach is technically satisfactory but conceptually awkward. Quantification – as induced by predicates of propositional

---

[24] See Pearson (2013) for a detailed empirical justification of this claim.

attitude – is on triples of worlds, times and individuals. We have clear evidence – as discussed above – that with a subclass of these predicates times can be shifted. *PC follows if we assume that temporal shifting triggers individual shifting.* This is a technical possibility and nothing prevents it from applying. However, it is not a technical necessity. It is perfectly conceivable, from a purely conceptual perspective, that temporal shifting does *not* trigger individual shifting. For a Martian language with no PC-effects, it would be just enough to assume that this is the case. In other words, it seems that, as things stand now, we have a correct technical description of the phenomenon, but we still do not have a solid conceptual justification for it. In particular, it seems to us that the basic ingredient that is still missing is that we have a solid empirical justification for why temporal extension takes place (based on the semantics of the relevant predicates) but no serious conceptual explanation for why *individual extension* should also be triggered in the semantics of these predicates. In more plain terms, given the analysis we have, if the data in (28) were different from how they are, we would really be surprised. However, if PC did not exist, we would have no reason to be particularly surprised: we would simply infer that temporal extension does not trigger individual extension. The obvious question is thus: Is there a way to achieve a higher level of explanatory adequacy? More particularly, why should we have a phenomenon of individual extension, paralleling temporal extension? We think that the semantics of theta-overwriting provides an explanatory answer to this important question.

Suppose that we adopt the results summarized at the end of section 5, including the conjecture that Control generally involves theta-overwriting, even for the cases (and the languages) where overwriting is not limited to the structures giving rise to IEM-readings. As already emphasized, one consequence will be that there is no need for PRO. This approach would also rule out Landau's analysis, based on agreement intervention effects caused by Infl-PRO agreement and Infl-to-C movement[25]. Consider now the effects of theta-overwriting. We have proposed that when the higher theta-role overwrites the lower theta-role, the latter is deleted for the purposes of the syntactic derivation and replaced by the former. This was tentatively rendered in (10), reproduced below as (29) for the reader's convenience:

(29) $\lambda x \lambda y \; [\text{Exp}_{V1}(x) \ldots \text{Ag}_{V2}(y)] \; \lambda x \; [\text{Exp}_{V1+V2}(x)]$

This is intended to express the fact that the two theta-roles are reduced to one

---

[25] See Landau (2000).

for the purposes of the syntactic computation, whilst the 'deleted' theta-role is still allowed to remain active at the interpretive interface. What does this mean exactly? In order to see this in more detail, let us consider the context in (30a), which triggers a PC-reading of the control structure in (30b):

(30)   a.   The problem was solved by the team, in fact everyone contributed something to the solution.
            b.   And actually, on that occasion, John enjoyed solving/liked (to) solve the problem as a team.

Under the analysis just sketched, the Experiencer role of the higher verb overwrites the Agent role of the lower verb in (30b). Roughly, this entails that syntax expresses a reading according to which there is no issue of referentially identifying who solved the problem: there is simply the experiencer of a mental state of joy that is also characterized by the feeling of solving the problem. The very point is now that in the context of (30a), this does not exclude that the referential content of the deleted Agent role trigger a process of pragmatic enrichment. If we put these two observations together, we get the following interpretation for the control structure in (30b): "John was the bearer of an experience of joy that involves the feeling of solving the problem, *and this feeling included the fact that more than one person was involved in the solving*".

At this point, there is a crucial observation to be made. Suppose that we stick to the context defined by (30a), while replacing the control structure in (30b) with the simple clause in (30c) below:

(30)   c.   *And actually, on that occasion, John solved the problem as a team.

The question is: Why is pragmatic enrichment (leading to individual extension effects) allowed in the complex structure in (30b) and ruled out in the simple structure in (30c)? Notice that this observation corresponds to the well-known fact that PC is a property of embedded clauses, and does not extend to root clauses.[26] We think that the answer to this question is quite straightforward within the framework that we have developed. In (30c) the Agent role is syntactically discharged as a semantically singular argument (i.e. John), whereas the predicate (i.e. 'solve the problem as a team') is semantically plural. All we need in order to derive the ungrammaticality of (30c) is the hypothesis that there cannot be pragmatic enrichment in these contexts of 'formal mismatch', in which the rules of projection have been violated (a syntactically expressed

---

[26] See Sheehan (2012) for an updated discussion, and the references cited therein.

semantically singular argument cannot be made compatible, in the pragmatics, with a semantically plural predicate). However, there is arguably no formal mismatch in (30b), i.e. in the case of the control structure. The reason is that the Agent role is not syntactically expressed in (30b): it has been overwritten by the Experiencer role of the higher verb, that is, it has been deleted *for the purposes of the syntactic computation*. If we accept this as a crucial ingredient of the semantics of overwriting, it is no longer the case that the process of pragmatic enrichment induced by the context in (30a) conflicts with the singular semantics of the Agent role. In fact, syntax does not encode a singular semantics of the Agent role, since the latter has been virtually suppressed *as far as syntax is concerned*. There is thus no formal mismatch to be repaired. At the same time, and crucially for our purposes here, *the Agent role remains active for the systems of interpretation*: pragmatic enrichment can thus felicitously take place in the context of (30a), yielding the contextually enforced interpretation according to which the feeling of joy is linked to a shared solution of the problem.

We propose that this is always the case for all control structures: PC simply corresponds to the application of a mechanism of pragmatic enrichment that re-activates the deleted theta-role in the pragmatics (after all, the relevant theta-role, though deleted in syntax, still remains an obligatory interpretive feature of the embedded predicate in itself). This entails that PC—as an effect of individual extension, as in Pearson (2013)—is a necessary ingredient of the semantics of Control: whenever theta-overwriting takes place (within the framework defined here, in all control configurations), pragmatic enrichment of the lower theta-role (hence PC-effects) are predicted to be possible. If this is so, there is a further important conceptual advantage. It has long been recognized that one of the 'mysteries' of PC is the fact that PC is not symmetric. Namely, it is possible for the referent of the lower theta-role to include the referent of the higher theta-role, but it is crosslinguistically excluded (hence a reasonable effect of the UG format) that the referent of the higher theta-role includes the referent of the lower theta-role (that is, 'inverse PC' is excluded). Given the line of analysis just sketched, there is nothing to explain here: inverse PC would be tantamount to assuming a mechanism of pragmatic enrichment of the referential content of the higher theta-role. However, this theta-role overwrites the other one, and remains thus syntactically active: if it is syntactically realized as a semantically singular argument, it is not allowed to agree with a semantically plural predicate. In other words, there simply is no issue of inverse PC in the present framework.

There are two final problems to be solved. As it stands now, our approach clearly over-generates. It is predicted that all control structures are compatible with PC, contrary to the facts. However, this simply means that we have to take into account the important results reached in (Pearson 2013): individual extension (that is, the possibility of pragmatic enrichment as a consequence of the semantics of theta-overwriting) is only possible for those predicates that allow temporal extension. For instance, *claim/believe/pretend* do not: it follows that PC is excluded as well. A possible objection against this line of analysis is that it does not improve on Pearson's analysis: there, it had to be stipulated that temporal extension involves individual extension, here it has to be stipulated that individual extension involves temporal extension. However, this criticism would be unfair. Remember that the real difficulty was to understand the conceptual motivations for individual extension. It is this problem that has been solved now: individual extension (hence PC) is nothing else than a mechanism of pragmatic enrichment of the lower theta-role that is made possible by the semantics of thematic overwriting and that is blocked in all other cases. On this basis, all we have to assume in order to eliminate over-generation is a quite reasonable *uniformity condition on variable extension*. This condition states that, once granted that time and individual extensions are independently motivated, they cannot apply independently of each other, that is, they cannot apply non-uniformly (asymmetrically). The behavior of canonical propositional attitudes that do not admit PC (like 'claim' and 'believe') immediately follows. These are cases where the uniformity condition is violated, since temporal extension is impossible (on independent grounds with respect to individual extension). Monothematic restructuring predicates also do not admit PC: in this case, there is a single theta-role involved, which undergoes syntactic realization, to the effect that pragmatic enrichment is excluded. In all residual cases (including the cases involving an IEM-reading), the lack of PC can only be due to a violation of the uniformity condition, i.e. to the impossibility of applying temporal extension. Empirically, this seems to be correct (Pearson 2013).

The second problem to be addressed is that it might not be immediately clear why theta-overwriting should be compatible with individual extensions through pragmatic enrichment. After all, the (original) semantics of overwriting is based on referential indistinguishability of the two theta-roles, and this might seem to be incompatible with the fact that the referential indexes associated to the two theta-roles end up as different, as a consequence of the extension effect that applies to one of them. Moreover, it is just referential indistinguishability that allowed us to derive the impossibility of proxy-readings with lexical reflexives. So, one might ask, why should overwriting rule out proxy-readings and rule in PC-effects (i.e. individual extension by means of pragmatic enrichment)?

The answer is in fact quite straightforward. Proxy-readings may be assumed to result from the application of Skolem-functions to entity-referring expressions. Based on (Reuland and Winter 2009), Skolem-functions are defined as in (31):

(31)   A function f of type (ee) with a relational parameter PR is a Skolem function if for every entity x: $PR(x, f_{PR}(x))$ holds.

Intuitively, Skolem-functions map an individual x into a referentially distinguishable individual y, though x and y are related to each other in terms of some relevant relational parameter, as formulated in (31). For instance, the wax-counterpart of John is a distinct object, though it is related to John in terms of the relation 'wax-copy'. The mechanism of individual extension that we have proposed for PC does not satisfy this condition of referential distinguishability. The reason is that this mechanism does not map a referential index i into a distinct referential index j, but it simply extends i by building up an index set $\{i, j, k, \ldots\}$ that contains the original index i. In this way, the intersection of the extended object and of the original object is by definition not empty, and the associated referential indexes turn out to be non-distinguishable. It follows that pragmatic enrichment is fully compatible with overwriting, since it does not violate the condition according to which overwriting is fed by referential indistinguishability. Conversely, proxy-readings violate this condition. To get the correct results, it suffices to interpret indistinguishability as in (32) below, an acceptable move on conceptual grounds:

(32)   Two theta-roles are referentially indistinguishable iff they either end up assigned to the same object or they end up assigned to objects that are in a part-whole relation to each other.

Let us conclude this section by exemplifying how IEM-effects and PC-effects are made compatible by means of an example. Take the continuations in (33b) and (33c), given the context introduced by (33a):

(33)   a.   John had been seeing Mary for a long time. Eventually ...
       b.   ... *he kissed.
       c.   ... he wanted to kiss.

We know what the reason is of this grammaticality contrast: pragmatic enrichment through individual extension is permitted only in (31c), as discussed above. What about the interpretation of (31c)? Remember that this is a sentence that involves both an IEM-reading and a PC-reading. Well, (31c) reports on a subjective experience of John's, more exactly on his willingness, which is immediately given to him, to be the experiencer in an event of kissing, whereby it is contextually determined that Mary also participates, as an Agent, to this very same event of kissing. There is nothing contradictory or awkward about this reading, which puts together PC-effects and IEM-effects. On the contrary, this seems a non-trivial elucidation of the semantics associated with (33c), as the result of a well-defined set of concomitant conditions.

7 Conclusions

In this contribution, we have proposed that the syntax of Control is based on a lexical operation of theta-overwriting, whereby a theta-role overwrites another theta-role. Originally, theta-overwriting arises as the semantics of a particular class of *de se* readings, namely the cases where *de se* gives rise to immunity to error through misidentification (IEM-readings). Since this is exactly the semantics associated with restructuring predicates that assign a subject theta-role, theta-overwriting constitutes a principled grammatical solution for what we have dubbed *Cinque's paradox*. This is the requirement that the higher subject theta-role be somehow syntactically expressed even if no VP is projected (hence no thematic subject position is available), since restructuring verbs are functional heads filling dedicated position within the functional hierarchy of the clause. We have proposed that theta-overwriting may extend beyond the original IEM-structures, as a matter of cross-linguistic variation. Finally, we have proposed that it is the semantics of theta-overwriting that explains the availability of Partial Control and some of its core properties, as for instance the non-existence of inverse PC and the impossibility for individual extension effects to apply in simple clauses. In order for PC to apply, two conditions have to be satisfied: first, there must be theta-overwriting; second, the extension must apply uniformly to all variables quantified over, and this reduces to the requirement that the semantics of the predicate also allow, on independent grounds, temporal extension, in the sense of Pearson (2013). All in all, we argued that a particular class of *de se* readings can contribute to a better understanding of control phenomena, including Partial Control. Some long-debated issues have been elucidated by means of an original combination of syntactic and interpretive ingredients (including conditions on language use), with an attempt at eliminating any residual stipulation.

References

Abusch D. (1997). "Sequence of tense and temporal *de re*". *Linguistics and Philosophy* 20(1): 1–50.
Boeckx C., N. Hornstein & J. Nunes (2010). *Control as movement*. Cambridge: Cambridge University Press.
Castaẽda H.-N. (1968). "On the logic of Attributions of Self-Knowledge to Others". *The Journal of Philosophy* 65(15): 439–456.
Chierchia G. (1989). "Structured meanings, thematic roles and control". In G. Chierchia & B. Partee (eds). *Properties, Types and Meanings II*. Dordrecht: Kluwer. 131–166.
Cinque G. (2004). "Restructuring and functional structure". Ms., University of Venice.
Delfitto D. & G. Fiorin (2008). "Towards an extension of *de se*/*de re* ambiguities: Person features and reflexivization". *Lingue e Linguaggio* 7: 25–46.
— (2014). "Control, attitudes *de se* and immunity to error through misidentification". *Rivista Internazionale di Filosofia e Psicologia* 5(2): 184–206.
— (2015). "A Case for Reference to Phenomenal Experience in Natural Language". *Cahiers de linguistique française* 32: 119–132.
Delfitto D., A. Reboul & G. Fiorin (to appear). *Immunity to error through misidentification and (direct and indirect) experience reports*, Berlin: Springer.
Gallagher S. (2000). "Philosophical Conceptions of the Self: Implications for Cognitive Science". *Trends in Cognitive Sciences* IV(1): 14–21.
Higginbotham J. (2003). "Remembering, imagining, and the first-person". In A. Barber (ed). *Epistemology of language*. Oxford: Oxford University Press. 496–533.
Hornstein N. (1999). "Movement and control". *Linguistic Inquiry* 30(1): 69-96.
Kaplan D. (1989). "Demonstratives". In J. Almog, J. Perry & H. Wettstein (eds). *Themes from Kaplan*. Oxford: Oxford University Press. 581–663.
Landau I. (2000). *Elements of Control: Structure and Meaning in Infinitival Constructions*. Dordrecht: Kluwer Academic Publishers.
— (2003). "Movement Out of Control". *Linguistic Inquiry* 34: 471–498.
— (2015). *A two-tiered theory of control*. Camridge MA: MIT Press.
Morgan D. (2012). "Immunity to error through misidentification: what does it tell us about the *de se*?". In S. Prosser & F. Recanati (eds). *Immunity to Error Through Misidentification: New essays*. Cambridge, Cambridge

University Press. 104–124.
Pearson H. (2013). *The sense of Self. Topics in the semantics of* de se *expressions*. PhD dissertation. Harvard University.
Pryor J. (1999). "Immunity to Error Through Misidentification". *Philosophical Topics* 26: 271–304.
Recanati F. (2007). *Perspectival Thought*. Oxford: Oxford University Press.
Reinhart T. (2002). "The Theta System – An Overview". *Theoretical Linguistics* 28: 229–290.
Reuland E. (2011). *Anaphora and language design*. Cambridge, MA: MIT Press.
Reuland E. & M. Marelj. (2013). "Clitic SE in Romance and Slavonic revisited". In I. Kor Chahine (ed). *Current Studies in Slavic Linguistics*. Amsterdam: John Benjamins. 75–88.
Reuland E. & Y. Winter. (2009). "Binding without identity: Towards a unified semantics for bound and exempt anaphors". In S. Devi, A. Branco, & R. Mitkov (eds). *Anaphora processing and applications*. Berlin: Springer. 69–79.
Rizzi L. (1978). "A Restructuring Rule in Italian Syntax". In S.J. Keyser (ed). *Recent Transformational Studies in European Languages*. Cambridge MA: MIT Press. 113–158.
— (1982). *Issues in Italian Syntax*. Dordrecht: Foris.
Shoemaker S. (1968). "Self-reference and self-awareness". *The Journal of Philosophy* 65: 555–567.
Wittgenstein L. (1958). *The Blue and Brown Books*. Oxford: Basil Blackwell.
Wurmbrand S. (2002). "Syntactic vs. semantic control". In C.J. Zwart, & A. Werner (eds). *Studies in Comparative Germanic Syntax: Proceedings from the 15th Workshop on Comparative Germanic Syntax*. Amsterdam: John Benjamins. 95–129.
Zubizarreta M.-L. (1982). *On the Relationship of the Lexicon to Syntax*. Ph.D. Dissertation, MIT.

Chapter 5
A Pragmatic Promenade in the French Landscape of Colours

Louis de Saussure

Abstract In this paper, we observe a number of surprising features of French colour terms and discuss them in relation to the Basic color terms theory (originally by Berlin & Kay 1969). We discuss the morphological combinations which should be compatible with basicity but are actually odd, like *orangeâtre* (orangeish) and, converserly, those which should be odd but are fine, like *beigeâtre* (beigeish). We also notice that some combinations which are normally blocked, like *azurâtre*, can nonetheless occur in very special contexts. We claim that these cases are actually possible if pragmatic conditions are met, however raising specific meanings. In particular, we suggest that the logical inconsistencies that occur in some of these combinations are overcome by transferring the scalarity entailed by the suffix *–âtre*, which is irrelevant if applied to a non-basic term (thus to a term which supposedly doesnt cover any chromatic range or scale), to a salient and relevant non-chromatic scale. Such processes of lexical enrichment are however more difficult with verbs of colour. Finally, we briefly discuss the status of derived basic terms like *orange* and *violet*, and of non-basic atypical words like *roux* and *olive* which have some features of basicity.

Keywords Color terms · Morphology · Pragmatics · French.

Berlin & Kay (1969/1991) list a number of criteria (without discussing them in details) that allow for the identification of basic colour terms. One of them (criterion V) is morphological: basic terms have a distinct morphological distribution from non-basic terms (however Berlin & Kay suggest that this criterion should be referred to in case of doubts). In IE (indo-european) languages such as English or French, basic terms allow for suffixations, whereas non-basic terms don't. Hence, they notice, a basic term like *red* produces *reddish*, whereas a non-basic term like *chartreuse* does not produce **chartreuseish*. The same holds in French: *rouge* produces *rougeâtre*, but *turquoise* doesn't allow **turquoiseâtre*.

It is interesting to observe that in French a similar distribution holds with verbal derivations having reflexive readings meaning *to become of colour X*. In English, some colour verbs use a suffix (*to redden*), while others are simply produced by transcategorisation (*to green*). French has a homogeneous derivation system: *rouge, vert, brun* etc. produce *rougir, verdir, brunir*, etc. whereas *sapin* or *outremer* will not allow **sapiner* or **outremerer*. This remark holds only for these verbs in their reflexive (therefore non agentive) interpretation (*becoming red / blue* etc.). Some non-basic colour verbs are perfectly natural with transitive meanings (*vermillonner* for *rendering X vermillion, dorer* for *covering with gold* for example). Reflexive basic colour verbs can also have transitive readings (*rougir* in the sense of actively making something red), but we are not looking at this meaning here.

There are a number of issues regarding this morphological criterion. First, despite their being listed in most dictionaries, at least one French basic term's derivation seems quite unnatural: *orangeâtre*[1]. *Orangeâtre* does appear occasionally but its frequency is very weak (168 occurrences on the Internet in one year, according to a raw Google.fr® research done in February 2015) compared to the derivation of the other basic terms (*jaunâtre* tops the list with 34000 occurences and *rosâtre* is at the very bottom with around 4000 occurrences. There are also some doubts regarding *violâtre*, which has 560 occurences in one year).

Second, we notice that verbal derivations with reflexive meaning are also unnatural with *orange*. The verb *oranger* works only with transitive meaning (in *le ciel orangé du couchant*, *orangé* is a denominal adjective). The verb *oranger* does not exist in French: there is no tensed phrase such as *Le ciel *orangea dans le couchant* (compare with *Le ciel rougit / rosit / jaunit* etc.

---

[1] See Bloemen & Tasmowski (1983); these authors present an analysis of the frequency of colour terms in French and discuss the relation between frequency and basicity, which they see rather as a continuum.

dans le couchant). A similar observation seems also to hold with *violet* and *marron* (the prevalent term for *brown* in the French of France—*marron* doesn't occur in Switzerland for example and it is not a basic term anyway, merely a very common substitute).

Third, we notice that some non-basic terms do allow for occasional derivations: *olive* allows the derivation *olivâtre* very naturally, but not *oliver. *Blond* doesn't allow *blondâtre (but *blondasse*, which is slightly different) but produces *blondir*. *Roux*—probably the most interesting of French colour terms—allows both *roussâtre* and *roussir*.

Fourth, we notice that some non-basic terms that do not standardly provide –*âtre* derivatives nonetheless have some potential for it. *Beigeâtre* appears 78 times in our list and is listed in some major dictionaries, and even the unlikely *azurâtre* is present in a few poetic texts.

We suggest that a pragmatic explanation could account for most of these facts.

Let us consider Berlin & Kay's (1969/1991) criterion V through the lens of pragmatics and conceptual (categorical) cognition. It is usually assumed that –*âtre* indicates approximation (Kleiber 2008); we stress however that the 'approximation' here is in fact relative to one of the available prototypes (focal points): a *rougeâtre* and a *verdâtre* are indeed red and green but remote from any of the prototypes of these colours. Thus X–*âtre* indicates remoteness not from X (which would amount to being not-X) but from the prototypes(s) of X (see de Saussure 2014 for an elaboration). The difficulty of applying the meaning function of –*âtre* to *orange* comes from the fact that *orange* does not clearly identify a chromatic range but rather a definite shade: since the task of identifying a type of colour that is still a sort of orange but one which is remote from a prototype of orange becomes cognitively very weird to perform and amounts to saying that some colour is 'approximately precisely' orange, or something of the like.

Indeed, *orange* doesn't encompass an easily accessible subordinate lexicon that clearly distinguishes between a variety of shades (except darkness or lightness of course): expressions like *orange abricot* or *orange feu* seem quite unnatural in French, contrary to *jaune citron, bleu nuit* or *rose bonbon*. As a consequence, *orange* seems to share properties of both basic terms (in particular it is not itself a subcategory of some overarching term) and of non-basic terms, since it doesn't have subcategories identified with lexemes and therefore is not really acknowledged conventionally as encompassing a chromatic range (regardless of the obvious fact that it does so, of course). Thus, *orange* seems to behave as if it were one and only one precise tone (whereas it is not in reality nor even in perception, needless to say). We lack the words in ordinary language for shades of orange, and this, we suggest, leads to a problem when it comes to combining with *–âtre*. Yet, we won't say that this explanation is entirely linguistic.

We suggest that there is actually a pragmatic mismatch in representing the approximation (by the suffix *-âtre*) of a specific tone in a single lexical categorization with non-basic terms.[2]

Sperber & Wilson (1986/1995) have nicely shown the range of reasons for which it is fair to assume that cognition is geared to the maximization of relevance, relevance being an equilibrium between cognitive effect (informativity) and processing cost (effort of decoding and drawing informative inferences). In the case of non-basic terms, that is, of precise shades within a chromatic range itself denoted by an abstract, basic, term, we suggest that the search for relevance is normally not successfully achieved by those morphemes as it would involve a notion of 'being approximately of a specific shade'. The reason for this is that it does not provide a piece of information significantly different or more meaningful than what would be achieved by the derived generic, basic, term. In other words: approximating a specific shade does not normally provide significant increase or difference in informativity relatively to the approximation of a generic basic colour. As an example, *sapinâtre is odd inasmuch as it is pragmatically redundant with the more abstract and generic *verdâtre* without providing an easily graspable distinct or richer meaning. This is probably sufficient to rule out the existence of such terms in the lexicon.

But we insist that this is pragmatic, not strictly linguistic: there is no logical-semantic inconsistency in being approximately turquoise, and wordings such as *approximativement turquoise* are perfectly fine. It is the concatenation of the notions in a single lexical unit that raises issues of relevance with regard to the generic counterpart.

This assumption relates to the distinction between information provided by single lexical items and periphrases or complex expressions. The rarity of a categorial need is certainly a factor influencing lexicalisation, since lexicalisation in turn provides a facilitated means of expression (and of reasoning). But there is far more to this than mere frequency of conceptual need: some constraints do actually apply to conventionalization which are due to the human cognitive system; an example of this phenomenon is the inability of languages to conventionalize the quantification *not-all* (see Newmeyer 2009), even though nothing prevents us to utter something like *Not all the students came to the party*, or to convey this quantification through scalar inferences:

---

[2] On the suffix *–âtre* and its semantic values, see Bottineau (2010).

Some students came to the party. We venture that a similar constraint applies to the morphological combinations that we observe not only in non-basic terms but also in *orangeâtre*, however on the lexical level, thus at the level of conceptual categories (not grammatical ones as in *not-all*). Since conceptual categories are flexible (Barsalou 1987), we suggest furthermore that the constraint here is pragmatic; as a consequence, such terms are not completely impossible: they simply need the appropriate context to make sense (be relevant). Therefore they are occasionally subject to contextual accommodation, provided that the hearer has ways to raise assumptions about specific intended interpretations that are apt to fulfil the expectations of relevance.

Getting back to the case of *beigeâtre*, our intuition is that the supplement of meaning is not simply a derogatory one (a supplement of meaning that this suffix bears commonly, but not obligatorily[3]); it has a particular flavour which is due to the way this derogatory meaning occurs. It is specifically achieved through a type of metarepresentation that bears some similarities with irony. Judging, for example, a sofa as *beigeâtre* will typically be interpreted as concerning a sofa which does not deserve to be called by the name of the shade *beige*. The utterance thus represents a subtle sort of mockery involving targeting someone (real or imaginary) who would call this sofa *beige*, which is judged ridicule, the conclusion being that the sofa is less-than-*beige* in aesthetic quality. The translation from approximation of colour to approximation of beauty or purity achieves relevance in this kind of cases.

The case of *beigeâtre* is however special in the sense that it has some degree of conventionalisation: dictionaries list the word as indicating 'unpleasant' or 'dirty' *beige*. Yet the same explanation makes also sense for cases without conventionalisation.

For example, some terms bear strong connotations of beauty and therefore cannot be accommodated towards not-beauty if derived by *–âtre*. Take *azurâtre*, which hardly occurs at all, but seldom appears in poetry or literary works. *Azur* bears very strong positive connotations (e.g. *Côte d'azur*). A poet writing about a *triste azur* would make an oxymoron of some sort, but *azurâtre* as a single lexical unit can hardly be oxymoronic on its own, since merging a full oxymoron in a single morphological construction is unlikely. Therefore, not only the 'approximation' meaning is pragmatically odd, for the reasons given above, but its accommodation by translation to the scale of connotations

---

[3] *Rougeâtre* and *jaunâtre* are often associated with beauty: "*Elle baissa vite, avec embarras, son bras nu, belle anse rougeâtre*" (Colette, *La naissance du jour*); "*champignon d'un beau jaunâtre*" (Secrétan, *Mycographie suisse*); similar remarks can be made about other derived basic terms, even though the derogatory connotation seems more common with them.

is not available. An *azurâtre* cannot be an unpleasant kind of *azur* because the connotations of *azur* block this interpretation. Therefore a poet mentioning a *brume azurâtre*[4] might tell us about a foggy day with touches of true and beautiful blue sky behind the moving screen of the mist, for example. It's more common in such cases to use the adjective *azuré*, but this one still carries a notion, symbolic or real, of agentivity (the French singer Serge Gainsbourg, in one of his famous songs, talks about the *ciel azuré*: that sky is beautiful as a painting).

What about *roussir* and *olivâtre*? *Olivâtre* tends to be specialized for faces, which implies that it tends to be a substitute of *vert* specifically for the shades signalling sickness. Non-basic substitutes of basic terms tend to behave like the term for which they stand, just as *blond*, but still keep some of their original non-basic properties, just as *blond* cannot produce **blondâtre* (*blond* is therefore the opposite of *olive* in this respect).

Roux is fascinating. It is very complicated to identify the kind of colour which it designates. It seems to be a mixture of brown and red, without being clearly a type of red. But it also qualifies the orangeish—if I dare say—or even the frankly orange type of hair ('raid hair'; the French novelist Jules Renard story *Poil de carotte* is about the childhood of a red-haired child). In this respect, *roux* looks like one of these basic terms with wide chromatic scope but unclear focal prototypes or maybe several prototypes in chromatic zones relatively remote from one another. It does also fit the same morphological distribution as basic terms in general, allowing *roussâtre* and *roussir*.

Does *roux* belong to the category of basic terms? It could be that *roux* fills a gap for borderline yellow-orange-red-brown colours, rather than a specific shade inside one of them; it is possible that *roux* gains some level of autonomy in the lexicon without having a clear focal point. According to the other criteria by Berlin & Kay (1969/1991), *roux* looks like a basic term inasmuch as it is monolexemic (criterion I), it is not a shade of another colour (criterion II), it is not specific for particular classes of objects (criterion III), but it is unclear as for criterion IV, which is about its saliency among speakers as a colour term. It is not the name of an object (criterion VI), it is not a borrowing (criterion VII) and has no morphological complexity (criterion VIII) (see also Kay et al 2010: 21 for a summary and reassessment of the criteria). The reason for which it would not be classified as *basic* lies only in the fact that it is not saliently a colour term for French speakers, which is a weak notion to decide for the belonging of a word to a certain category. Much clearer is the fact that it has controversial focal points. A nice experiment still to design but easy to

---

[4]in Alain van Crugten, *Personnes déplacées*.

perform would consist in asking subjects to pick-up focal points for *roux* in the Munsell table and do the reverse with other subjects (i.e. to name those focal points). It's predictable that they will fall into another category, like *orange*, *red*, *brown*, even though maybe at their boundary. This only would probably suffice to show that *roux* is in some sense an outsider among colour terms: basic, but composite.

Recent work (besides the huge research of the *World color survey* directed by Paul Kay) have shown ways to think differently about colours than through a simplistic *all-universal versus all-relativist* opposition where holding one position prevents from whatever coming from the other side. Certainly, a number of scholars have been trapped in this polarity (and some still are), but not only recent experiments have shown how the stabilization in the lexicon of basic colour terms influences perception (typically, a lexical boundary will induce a distortion of the perception of actual distances between shades, see Gilbert et al. 2006 and for a survey Regier & Kay 2009 and more recently Reboul 2015), but new approaches, in particular by Jraissati (2009), and new ways to think about perception and language in the domain of colours (Reboul 2015, Ciaccio 2015), open to more elaborated models where language adds categorical tools without erasing fundamental abilities. What is more, as Jraissati (2009) suggests, there is some notion of degree in the 'basicity' of colour terms. We venture that abstraction, which is the main feature of basic terms, is implemented in languages gradually and to various degrees across time, and therefore more recent terms have chances to behave half-way between basic and non basic categories. *Violet* might be a case of this sort (Jraissati 2009). *Orange*, our data suggests, is an even clearer one. *Violet* appears in French as a colour term in XIIIth century French, and *orange* only around 1550 (Mollard-Desfours 2008).

Roux, again, resembles *orange* as it is without hyponyms (there is no such thing as *roux renard* for example), but still allows far better that *orange* the derivations discussed above. The problem with *roux* might be that it retains still nowadays a notion of 'burning' (a number of expressions in contemporary French still match *roux* with fire, as in *ça sent le roussi*) ; and what is burnt becomes of a colour that depends on the material burnt.

Roux is however not the only colour term that traces back to fire in a way or another. Looking at various IE roots, it's quite clear that a number of terms for *red* originated as a separation from *black-dark* (it's particularly clear in Slavic), and a number of terms for black such as latin *ater* derive from an older notion of fire or smoke. Conversely, old Germanic *blakaz*, which will provide not only *blank* (shining), but also French *blanc* and English *black* (which replaced *swaert* in middle English; *blac* could designate both a shining white and a shining ink in old English) trace back to a notion of fire. The fire is of changing colour, brilliant, red, yellow, but also making dark and obscure fumes. It is not extravagant to speculate that when colours started to differentiate in IE languages, a process not yet achieved in proto-indo-europan (which had roots for yellow-green, but not for blue and the susequent series), they did so by abstracting from natural categories, and typically from the fire, as far as the first one, historically speaking, is concerned (red). Note also, incidentally, that *blakaz* shares common origins with the range of *bl–* colours anchoring on an old PIE root with *bl*: *blue*, *blond* but also *flammeus* (one of the Latin non-basic terms for *red*) and the Slavic terms for white as in *Belgrade*, the 'white city'.

All this takes us too far away from language and cognition. The linguist however can't help walking in these landscapes where language meets nature and culture at the same time, certainly not explaining them and not entirely explained by any of them. Looking at small facts like this one unveils a little bit of the complex ways in which human history and histories went in order for individuals to grasp the outer world with verbal and cognitive tools, which in turn play a central role also in the development of cultures. As Reboul (2015) demonstrates: "language adds an entirely new cognitive dimension (...), but does not thereby alter the original abilities and representations". Far from being separated from one another by some ineffability of our relative worldviews, as a strong Sapir-Whorf view would romantically say, we are linked to one another by our common cognitive apparatus, which expresses itself in an unlimited number of ways but never without connection to our common human nature—which involves seeing the world in colours.

References

Barsalou L.W. (1987). "The instability of graded structure: implications for the nature of concepts". In U. Neisser (ed). *Concepts and Conceptual Development*. New York: Cambridge University Press. 101–140.

Berlin B. & P. Kay (1969). *Basic color terms.* Berkeley: University of California Press (2nd ed.: 1991).

Bloemen J. & L. Tasmowski (1983). "Les noms de couleur en français: Catégories et focus". *Linguistica Antwerpiensia* 17: 16–17.

Bottineau T. (2010). "Les valeurs sémantiques du suffixe -âtre, marqueur d'opérations sur le plan notionnel". *Syntaxe et sémantique* 11: 35–54.

Ciaccio L.A. (2015). "Color terms and color perception. Reconciling universalism and relativism". *Rivista Italiana di Filosofia del Linguaggio* 9(2): 1–13.

de Saussure L. (2014). "Remarques sur la distribution morphologique des termes basiques de couleur en français". *Travaux de linguistique* 69: 77–90.

Gilbert A.L. et al. (2006). "Whorf hypothesis is supported in the right visual field but not the left". *Proceedings of the National Academy of Sciences* 103: 489–494.

Jraissati Y. (2009). *Couleurs, culture et cognition. Examen épistémologique de la théorie des termes basiques de couleur.* ENS-EHESS Institut Jean Nicod, PhD dissertation, 316p.

Kay P., B. Berlin, L. Maffi, W.R. Merrifield & R. Cook (2010). *The world color survey*, Stanford: CSLI publications.

Kleiber G. (2008). "Adjectifs de couleur et gradation: Une énigme très colorée". *Travaux de Linguistique* 55: 9–44.

Mollard-Desfour A. (2008). "Les mots de couleur : des passages entre les langues et les cultures". *Synergies Italie* 4: 23–32.

Newmeyer F. (2009). "Peut-on reconstruire la langue des premiers êtres humains ?" *Nouveaux cahiers de linguistique française* 29: 99–113.

Regier T. & P. Kay (2009). "Language, thought, and color: Whorf was half right". *Trends in Cognitive Sciences 13*(10): 439–46.

Reboul A. (2015). "A new look on the Sapir-Whorf hypothesis on colours, based on neuroscientific data". In V. Bogushevskaya & E. Colla (eds). *Thinking colours. Perception, translation and representation*. Newcastle upon Tyne: Cambridge Scholars Publishing. 2–16.

Sperber D. & D. Wilson (1986). *Relevance. Communication and Cognition.* London: Blackwell (2nd ed.: 1995).

Chapter 6

How Logical is Natural Language Conjunction? An Experimental Investigation of the French Conjunction *et* *†

Joanna Blochowiak & Thomas Castelain

Abstract The aim of this paper is to study experimentally the role of structure and meaning in the processing of sentences linked by the conjunction *and*. Traditionally, different interpretations of conjunction in natural language are explained from a pragmatic perspective as the result of the interaction between its logical meaning and pragmatic principles governing the discourse. According to Grice and many of his followers, the semantics of *and* is equivalent to its logical, truth-functional definition, and the additional (e.g. temporal) connotations are pragmatically derived as implicatures. Quite recently, an alternative explanation has emerged from a syntactic perspective, according to which there is a structural ambiguity between the symmetrical (logical)

---

*The present paper is dedicated to Anne Reboul in celebration of her 60th birthday. Her extraordinary work, as well as her ever-supportive and gentle advice, has never ceased to inspire and encourage the authors of this paper in their own research.

†The present study was a part of Project N⁰ 100012_146093: "LogPrag: the semantics and pragmatics of logical words (negation, connectives, quantifiers)" carried out at the University of Geneva (2014-2017) and funded by the Swiss National Science Foundation. The first author is grateful to this institution for its generous financial and administrative support. The two authors wish to express their gratitude to Jacques Moeschler for his valuable collaboration on an earlier version of this paper.

and asymmetrical (temporal and causal) uses of conjunction. These approaches make different predictions in terms of the processing cost involved in interpreting utterances with *and*. This paper presents a reading time experiment conducted on the French conjunction *et* (and), which aims to test these predictions. In general, our results showed that the reading times of logical, temporal and causal interpretations increased in that order. In particular, we found that the difference between logical and causal interpretations was statistically significant, while the difference between logical and temporal was not. In the discussion, we examine the hypothesis that this lack of difference may be due to the difficulty in differentiating between logical and temporal interpretations inherent to natural language.

Keywords Logical words · Conjunction · Relevance Theory · Reading time experiment

1 Introduction

It is well known that the conjunction *and* can have various interpretations in natural language, such as logical, temporal or causal interpretations. Assuming structural uniformity among these different interpretations, semantic and pragmatic theories explain these interpretative differences by the general pragmatic principles of communication (Grice 1975, 1989; Posner 1980; Schmerling 1975; Carston 1993; Blakemore & Carston 1999; Blakemore & Carston 2005). However, quite recently, Bjorkman (2010) has put forth some arguments in favour of a structural division between these different interpretations. In brief, asymmetric interpretations (temporal and causal) in their syntactic structure involve a coordination of temporal phrases (TP), while symmetric interpretations (logical) imply a coordination of complementizer phrases (CP).

This paper presents an experimental study aiming to examine which elements prevail in the on-line processing of sentences with the French conjunction *et*. If structural considerations prove to have a cost in the processing of *et*, we could hypothesize a clear difference between the processing of asymmetric (temporal and causal) and symmetric (logical) interpretations. The latter should have shorter processing times, because they are composed of smaller syntactic structures (TP), whereas the former should exhibit longer processing times, because they are composed of larger syntactic structures (CP). In contrast, if pragmatic principles play a dominant role in the processing of sentences with *et*, we can expect the logical interpretation of *et* to be processed faster, since it constitutes the basic semantics of *et*, whereas the two other types of interpretation should be processed at a slower rate.[1] The determination of the "rapidity" of processing between temporal and causal *et* is to be found in different pragmatic theories. For instance, for Levinson (1983, 2000), the temporal interpretation should be quicker to process than the causal; however, this is not necessarily the case for Relevance Theory (Sperber & Wilson 1986/1995). The speaker interprets utterances based on her encyclopedic knowledge, which contains various mental schemes that allow the correct utterance interpretation to be reached. We will discuss both approaches in more detail in section 2.1.

2 Background

According to traditional pragmatic theories, the semantics of *and* is rooted in the meaning of the operator of the conjunction ∧ from propositional logic, which is minimal, in the sense that it takes into account only the truth-values of the conjoined propositions. Hence, the resulting complex proposition (p ∧ q) is true only if both the propositions (p, q) that compose it are true. Consequently, the logical operator of conjunction has the property of symmetry, which means that the result of its application is independent of the order of its arguments, i.e. p ∧ q is equivalent to q ∧ p. Table 6.1 gives the truth-table for ∧, where 1 = true and 0 = false.

p	q	$p \wedge q$
1	1	1
1	0	0
0	1	0
0	0	0

Table 6.1: Truth-table for logical conjunction

Strictly speaking, the conjunction *and* in logic is a binary operator inducing the logical relation of conjunction, in the sense that any two-place operation defines a binary relation when the value of the result of this operation is fixed (for logical operators, the value is fixed to 1). Here, if we assume that the complex conjunctive proposition is true, we arrive at a relation between the two conjuncts which holds if and only if both are true. Hence, this is a purely logical relation which applies to every conjunctive statement.

---

[1] However, we should emphasize that this traditional view has recently been questioned. Some studies on the topic of implicatures suggest that only scalar implicatures with lexical replacement can incur a processing cost. For more detail, see Tiel & Schaeken 2017.

However, it is well known that the spectrum of relations which can be expressed by *and* in natural language is very broad, and the nature of some of these relations makes the property of symmetry unsuitable. Furthermore, in some situations, the inverse property—asymmetry—is applicable, as is true for many temporal and causal interpretations.

The standard assumption within pragmatics is that the semantic nucleus of *and* is constituted by its logical meaning, and the hearer arrives at all the other interpretations via the pragmatic type of inference, with the help of pragmatic rules and principles of conversation. So, as far as the logical form is concerned, examples (1)-(3) are symmetric, and other layers of meaning come from some sort of pragmatic inference (basically implicature or explicature, depending on the pragmatic theory, which we will detail in the next section).

(1)   It is raining and it is windy. (logical interpretation)

(2)   John woke up and he took his shower. (temporal interpretation)

(3)   Mary pushed Max and he fell down. (causal interpretation)

In the remainder of this section, we will present in more detail three pragmatic accounts (section 2.1) and a syntactic approach to conjunction (section 2.2).

2.1   Pragmatic perspective

Grice's original proposal was formulated in reaction to ordinary-language philosophers of his time, for whom there was a fundamental gap between the meaning of logical words—i.e. operators, connectives and quantifiers from propositional calculus—and their natural language counterparts:

> It is a commonplace of philosophical logic that there are, or appear to be, divergences in meaning between, on the one hand, at least some of what I shall call the FORMAL devices—$\neg, \wedge, \vee, \supset, (x), \exists(x), \iota x$ (when these are given a standard two-valued interpretation)—and, on the other, what are taken to be their analogues or counterparts in natural language—such expressions as not, and, or, if, all, some (or at least one), the.
>
> [...]
>
> I wish, rather, to maintain that the common assumption of the contestants that the divergences do in fact exist is (broadly speaking) a common mistake, and that the mistake arises from an inadequate

attention to the nature and importance of the conditions governing conversation. (Grice 1975: 41–43)

In other words, according to a traditional philosophical view, a word such as *and* and its logical counterpart ∧ have two distinct meanings, with rather few links between them. For Grice (1975), this view was mistaken, because it didn't take into account the rules and principles guiding natural language conversation. For instance, in the Gricean example (4), below, *and* keeps its logical meaning, and its temporal value corresponds to an implicature—that is, it is inferred by way of general pragmatic principles which govern communication.

(4)   John took off his boots and went to bed.

In particular, the interlocutor here supposes that the speaker obeys the manner maxim of orderliness, according to which events are usually narrated in the order in which they happened.

However, the application of the maxim of orderliness does not elucidate the issue of causal interpretations, and Grice provided no solution to this matter. Nonetheless, post- and neo-Gricean approaches have offered some explanations for this phenomenon. We will focus on the approaches of Levinson and Relevance Theory, in that order, as they allow for the formulation of experimentally testable predictions.

Initially, Levinson (1983: 146) came up with an incremental algorithm, reproduced in (5), which was designed to calculate the various interpretations of *and*.

(5)   Given P and Q, try to interpreting it as:
    a.   "P and then Q"; if successful try:
    b.   "P and therefore Q"; if successful try also:
    c.   "P, and P is the cause of Q".

So, Levinson's 1983 proposal predicts that after the logical interpretation comes the temporal interpretation, followed by the causal interpretation.

A later development of Levinson's theory (Levinson 2000) shifted the focus onto the default type of interpretations. He proposed a theory of I-implicatures, which are generalized conversational implicatures triggered and guided by stereotypical information associated with the situations described by utterances. I-implicatures, which are pragmatically enriched meanings, automatically arise by default unless some specific contextual information prevails over them. In contrast, particularized implicatures are not automatic, and are triggered in specific contexts. Temporal and causal interpretations of *and* are typical examples of I-implicatures; temporal interpretation in particular is claimed to be the default one. When two past-tense event descriptions are conjoined with *and*, temporal sequential interpretation is the first to arise (6). However, since I-implicatures are pragmatic types of inference, they can be cancelled by the meaning of temporal adverbs (7), our knowledge of the world (8), or explicitly (9).

(6) Max went to the railway station and bought a ticket.
(Implicature: Max went to the railway station and then he bought a ticket.)

(7) Max went to the railway station and he bought his ticket before.
(The temporal implicature is cancelled by the interpretation of the adverb *before*.)

(8) Max went to the railway station and he bought his ticket online at home.
(The temporal implicature is cancelled by our world knowledge.)

(9) Max went to the railway station and he bought his ticket, but not in this order.
(The temporal implicature is cancelled by explicit denial.)

As far as cognitive processing is concerned, the updated version of Levinson's theory—and the theories defending default interpretations in general—predict that pragmatic enriched meanings come first by default, while semantic readings need a second stage to cancel default interpretations. The second stage is effortful and time-consuming, and semantic interpretations will therefore take longer to process than pragmatic ones. As such, in the case of the conjunction, the default type of theory predicts that temporal and causal readings of *and* (i.e. pragmatic interpretations) will be processed faster than symmetric/logical ones (i.e. semantic interpretations). We think that Levinson's 2000 proposal can deliver an even finer-grained prediction, which is that a temporal sequential interpretation (i.e. two conjoined past-tense event descriptions) will arise first, by default, followed by a causal one, if at all plausible; finally, a semantic/logical interpretation will appear whenever none of the pragmatic interpretations (temporal or causal) can be derived by the hearer.

Relevance Theory proposes a different solution (Sperber & Wilson 1986/1995). First of all, contrary to Grice and neo-Griceans like Levinson, neither temporal nor causal interpretations are considered to be implicatures, but are instead aspects of *what is said*, determined in an inferential manner (Carston 1988, 2002). In relevance theoretical terms, they are explicatures.

This account is motivated by many authors' observation (Carston 1988, 2002, Wilson & Sperber 1993, Moeschler 2000, 2010) that temporal and causal interpretations are elements of the meaning which contribute to the proposition expressed by utterances of *and*-sentences, as the classic examples borrowed from Carston (2002: 227) illustrate.

(10) a. Either he left her and she took to the bottle or she took to the bottle and he left her.
 b. He didn't go to a bank and steal some money; he stole some money and went to the bank.

As Carston points out, the embedding of the *and*-conjunction under the scope of logical operators, such as disjunction (10a) or negation (10b), demonstrates that the causal and temporal relations have to be part of the proposition expressed by the utterance of an *and*-sentence, as (10a) would otherwise be a mere redundant repetition of the form 'Either P or P', and (10b) would be a contradiction in the form 'Not P; P'. In other words, the temporal and causal relations need to be inferred by the hearer in order to arrive at the full propositional form evaluable in terms of truth and falsity. The point made by the relevance-theoretic account is that temporal and causal inferences, although pragmatic in nature, serve to enrich the propositional form of the utterance of an *and*-sentence.

How do these inferences come about? In the relevance-theoretic framework, the temporal and causal interpretations depend on the degree of accessibility of contextual assumptions that activate different mental schemas—for instance, of temporal or causal nature. The existence of such cognitive scripts or schemas, among other aspects of interlocutors' encyclopedic knowledge, ensures the adequate interpretation of a given utterance of an *and*-sentence. Such schemas capture types of causal or temporal situations which refer to frequently encountered sequences of particular events, actions or processes. For example, the correct assignment of the truth-conditions for a causal sentence, as in (11), is guaranteed by the interaction of the principle of relevance and some contextual assumption stored within a mental script concerning a type of described causal scenario.

(11) Mary dropped the vase on the tiles.

In this case, all the participants in the conversation have in their background knowledge a causal schema saying that a vase dropped on tiles will break. These contextual assumptions are made accessible precisely because of the encyclopedic knowledge that interlocutors share about this type of event.

As long as the contextual assumptions provide interpretations in accordance with the principle of relevance, other possible interpretations are ruled out.[2]

2.2 Syntactic perspective

Bjorkman (2010) presents an approach based on the analysis of embedded clauses, according to which there is a syntactic difference between symmetric and asymmetric coordination. The important point is that, in the embedded contexts, asymmetric interpretations are only available for the coordination of temporal phrases (TP), while the symmetric interpretations are accessible for the coordination of complementizer phrases (CP). Consider the following example, from Bjorkman (2010).

(12)   a.   The newspaper reported that a new mayor was elected and there was a riot.
       b.   The newspaper reported that a new mayor was elected and that there was a riot.

The structural difference between the two examples is clear. In (12a), the complementizer *that* is present once, which suggests that the conjunction *and* relates two structures of the type TP. *That* appears twice in (12b), implying the presence of a structure larger than TP, the CP structure. (13) illustrates the difference between the two structures schematically (Bjorkman 2010).

(13)   a.   ... *reported [$_{CP}$ that [$_{TP}$...] and [$_{TP}$...]]*
       b.   ... *reported [$_{CP}$ that ...] and [$_{CP}$ that ...]*

This structural difference is reflected in interpretative differentiation. When the TP-level coordination is involved, asymmetric interpretations are among those possible, while only the symmetric relations can appear in the case of the CP-level coordination. For instance, (12a) tends towards an interpretation in which the speaker observes a causal link between the two events described in the embedded clauses; according to this interpretation, the riot was causally linked to the election of the mayor. (12b), on the other hand, suggests no such interpretation; the two events described by the embedded clauses are instead processed in an independent manner, with no causal relation between them.

As such, the generalization proposed by Bjorkman (2010) is the following: the coordination of embedded clauses of the TP type provides interpretations with asymmetric relations (temporal or causal), whereas the coordi-

---

[2]The importance of causal schemas under the form of causal laws is also put forward in the Relevance Nomological Model (Blochowiak 2014, 2016).

nation of CP structures gives symmetric interpretations (logical). Assuming that the interpretation process is sensitive to the size of the language structures, Bjorkman's hypothesis predicts that symmetric interpretations, having larger structures, have longer processing times than asymmetric interpretations whose structures are smaller.

Below, we will look at previous experimental studies which aimed to test the predictions of different theories concerning the processing of logical words in general (Section 3.1), including the conjunction *and*. In Section 3.2, we will describe in more detail a study specifically designed to verify pragmatic and syntactic predictions regarding the conjunction *and* in English (Thompson et al. 2011, 2012). Finally, we will present our study on the French conjunction *et*, which was designed to verify the results obtained for English by Thompson et al. (2011, 2012).

3 Experimental investigation of logical words

The interpretation of logical words in natural language—and the particular question of how much logic there is in the natural language usage of logical words—has been under theoretical debate for a long time. Are we really sensitive to the logical interpretations as they are defined in classical logic? Or are these definitions purely theoretical artefacts with no cognitive reality whatsoever, implying that all we capture in our interpretation of the natural language expressions is their pragmatics? With the advent of experimental investigation of this subject, many answers to these issues have been found.

3.1 Logical words in children and adults cross-linguistically

In general, it seems that people are sensitive to the purely logical meanings of logical words. Numerous psycholinguistic studies have shown that children tend to interpret logical words as they are defined in classical logic (Noveck 2001; Papafragou & Musolino 2003; Guasti et al. 2005; Pouscoulous et al. 2007). These findings have been reported for several languages, such as French, English, Greek, Italian and German. A recent study on the acquisition of logical connectives in Chinese children confirms that the logical senses are the first to be acquired (Su 2014).

By now, it is well established (see Noveck & Reboul 2008 for an overview) that the 'non-enriched' semantic readings also require less effort to be processed by adults; this has been measured with various techniques, such as sentence processing tasks (for instance, Breheny et al. 2006 for *or*) or electroencephalography (for example, Nieuwland et al. 2010 for quantifiers).

Regarding quantifiers, there is abundant literature aiming to determine how the meaning of words such as *some* is understood (Horn 1972, Levinson 2000). Typically, it is assumed that *some* is a scalar term which is semantically compatible with *all*, but pragmatically enriched to *some but not all*. According to Levinson's (2000) approach, *some*—like *and*—has a default interpretation, *some but not all*. However, the experimental investigation does not confirm the predictions following from the default type of accounts, according to which semantic reading should take longer as a result of the stage cancelling the pragmatic default interpretation. For instance, in a sentence evaluation study, Bott and Noveck (2004) found that participants responded equally quickly to true underinformative sentences (semantic readings), such as *Some goats are mammals*, as they did to control items which were true or false statements, such as *Some mammals are goats* and *All goats are insects*. However, the participants took significantly longer to evaluate underinformative sentences which received false responses.

The tendency for rapid cognitive processing of non-enriched semantic readings has also been confirmed for disjunction and conjunction. The semantic meaning of disjunction is usually equated with the inclusive definition of disjunction in logic (that is, *one or both*) while the pragmatic enriched meaning is narrowed down to its exclusive interpretation (i.e. *or but not both*). In a sentence processing experiment, Breheny et al. (2006) presented participants with two types of contexts in which the same disjunctive sentences could be best interpreted as pragmatically enriched (upper-bound context, as in (14)) or not so (lower-bound context, as in (15)). During the self-paced reading task, the participants had to read the sentences chunk by chunk, by hitting the space bar (indicated by a slash in the examples below).

(14) Upper-bound context
While Mary and John were out shopping, /it started raining./John would get wet./Even though she did not have a lot of money,/she offered to buy him/an umbrella or a coat./

(15) Lower-bound context
It was highly probable that it would rain./Mary advised John/to dress accordingly./To avoid getting wet,/she suggested to him/to take with him/an umbrella or a coat./

The results are in accordance with previous findings showing that the participants took significantly less time to read the phrases containing *or* under the non-enriched semantic interpretation (lower-bound context) than the phrases with pragmatically enriched *or* (upper-bound context).

As we have seen earlier, the semantic non-enriched meaning of *and* in natural language is usually equated with the logical definition of conjunction, while the pragmatically enriched readings—i.e. temporal and causal interpretations—are derived via certain pragmatic mechanisms. Depending on the theory, these pragmatic inferences are claimed to be implicatures (Grice 1989, Levinson 2000) or explicatures—that is, the pragmatically inferred aspects of *what is said* (Carston 1993, 2002). The pragmatic meaning is usually claimed to arise later (in development or processing), except by default accounts such as Levinson's, in which pragmatic readings come first. Here again, default approaches do not receive experimental confirmation. For instance, Noveck et al. (2009) provided experimental evidence that pragmatic enrichments are cognitively costlier, and thus acquired later. In their experiment, participants first saw a small cartoon presenting two events in a certain order. They were then asked to answer questions formulated with the conjunction *and*, where the two events were presented in the reverse order. For instance, one question asked "Did Guillaume eat dinner at a friend's and pick up a cat into his arms?", while the cartoon presented the events in the reverse order. Children answered *yes* to such questions more frequently than adults, demonstrating once more their 'preference' for the semantic and logical interpretations of logical terms.

In the next section, we will present a reading time experiment on conjunction in English which seems to question these results. This study took into account predictions from not only pragmatic theories but also the syntactic account of conjunction reported in Section 2.2.

3.2 An experimental study on *and*

Bjorkman's hypothesis has been tested experimentally by Thompson et al. (2011, 2012). The experiment considered the processing time of sentences interpreted as semantically distinct (logical, temporal and causal) versus sentences interpreted as structurally distinct (asymmetric and symmetric).

The experiment was conducted with RSVP methodology (Rapid Serial Visual Presentation) (Forster 1970). This method presents the participants with sentences (word by word), in the centre of a computer screen, at a fixed rate. After the presentation of a sequence of words, participants are required to read aloud the sentence they have seen. Therefore, the measure of this experiment corresponds to the total production time of the sentences presented.

Their results seem partially to confirm the syntactic thesis formulated by Bjorkman (2010). Globally, the sentences implying symmetric interpretations were produced at a slower rate than those with asymmetric interpretations. However, the difference between logical and temporal interpretations is not statistically significant: it is only the difference between the logical and the causal *and* which meets this threshold.

Before going further, it is worth examining the protocol used by Thompson et al. (2011, 2012). First of all, the size of the subject group in their study was quite restricted, as it contained only eight participants. In addition, one should ask whether the RSVP methodology used—and, in particular, the measure of the production time—is best adapted for a study concerning the interpretation of *and*.

Even more problematic are the stimuli used in the experiment. First, certain sequences cannot be judged as purely temporal. The examples provided in (16) could very well be interpreted as causal *and*, but are classified as temporal *and*.

(16)   a.   The player scored and the team won the game.
            b.   The man fell and the woman laughed.
            c.   She won the lottery and they bought a yacht.

Moreover, the set containing stimuli with logical *and* is not uniform, because it includes episodic (17a), habitual (17b) and generic (17c) sentences.

(17)   a.   Gabriel ordered the pasta and Lily had some chicken.
            b.   Sarah studies in the library and Connie works from home.
            c.   Wolves hunt in packs and lions run in prides.

The lack of clarity in the temporal examples, as well as the non-uniformity in the set of logical examples, might have affected the production times, and as a consequence the final results of the experiment conducted by Thompson et al. (2011, 2012).

In order to verify the results obtained by Thompson et al. (2011, 2012), we developed a pilot experiment on the French conjunction *et*, which takes into account the problematic points detected in the choice of stimuli, and relies on a different methodology (Blochowiak et al. 2015). In the study presented in this contribution, we used the same procedure, but with a significantly larger number of participants, and included only native French speakers.

4 An experimental investigation of *et*

Outlined above were two types of approach which aim to explain different interpretations of *et*. The pragmatic approach predicts that the processing of the logical *et* is the least costly, followed by the temporal and the causal *et* (Levinson 1983), or by the causal and temporal *et* ordered according to the accessibility of contextual premises (Relevance Theory). In contrast, according to the syntactic approach, the asymmetric structures (temporal and causal *et*) should both be processed faster than the heavier symmetric structures (logical *et*).

To shed light on this question, we considered the three types of *et*, as in the experiment carried out by Thompson et al. (2011, 2012), trying to avoid problematic stimuli as much as we could. In particular, in the construction of our examples, we were careful to keep the verbal tense used in the sentences constant, in order to construct a stimuli dataset as uniform as possible. We used the French past tense *passé composé* across the three conditions, as it has one particular characteristic: unlike another French past tense, the *passé simple*, the *passé composé* does not impose a temporal sequential reading on sequences of sentences. For instance, the events described in (18) are understood to take place one after another because of the *passé simple* (PS), while in (19) the order is not imposed by the instructions related to the *passé composé* (PC).

(18)   Marie prépara le café. Les enfants vinrent à la maison.
         Marie prepared.PS coffee. Children came.PS home.

(19)   Marie a préparé le café. Les enfants sont venus à la maison.
         Marie prepared.PC coffee. Children came.PC home.

If hearers arrive at a temporal sequential reading of (19), it is instead due to Grice's maxim of orderliness (see Moeschler 2002, de Saussure 2003, Grisot 2015 for more detail on the French verbal system, and Grisot & Blochowiak (under revision) for experimental investigation of the interplay between French PS and PC, lexical aspect and temporal connectives).

4.1 A self-paced reading experiment on *et*

4.1.1 Material and method

Participants   Sixty-six native French speakers (45 females, Mage = 23.27, SD = 5.09, [18-46 y.o.]), students in Humanities at the University of Geneva, participated in the experiment. The participants were randomly assigned to one of the two experiments (45 to the Experimental condition and 21 to the Control condition).

Design In this study, we aimed to measure the reading times of complex sentences *P et Q*, depending on the three types of possible interpretations of *et*: logical, temporal and causal.

The study is based on two complementary experiments: the 'control' and 'test' experiments. The same sentences were used in the two experiments, the difference being that the comma was used in the control experiment to replace the conjunction *et*. Thus, the 'test' complex sentence *(P et Q) Il a neigé toute la nuit et les autoroutes sont impraticables* (It was snowing all night long and the highways are impassable) corresponds to the 'control' sentences *(P, Q) Il a neigé toute la nuit, les autoroutes sont impraticables*. The goal of the control experiment is to determine the mean reading time of *Q* without the presence of the conjunction *et*. The test experiment serves to determine the effect of the conjunction *et* on the mean reading time of *Q* as a function of the experimental condition (logical, causal, temporal).

To hide the aim of the experiment, the instruction given to participants was to judge whether the sequences of sentences were plausible or not. For example, the sequence *Marie a préparé les crêpes et Jean a passé l'aspirateur* (Marie made the pancakes and Jean did the vacuuming) should be judged as plausible by participants, whereas the sequence *Pierre est parti à la montagne et il a vu des extraterrestres en pyjama* (Pierre went to the mountains and he saw extraterrestrials in pyjamas) as implausible. The participants were also asked to answer as fast and as accurately as possible.

The sentences were constructed according to three criteria: a) all the sentences following the conjunction (*P et Q*) are in the *passé composé*; b) all the sentences are made up of common words; c) the number of syllables of the second sentence (*Q*) is between nine and twelve. Table 6.2 presents a sample of each category.

Procedure The experiment began with reading instructions, followed by a training phase (9 sequences) and an experimental phase (28 sequences with 18 test-sequences / 18 control-sequences). For each trial, a first sentence (*P*) appeared in the centre of the computer screen, and the participant had to press the space bar to move to the next sentence (*et Q / Q*). After reading the whole sequence (*P et Q* or *P, Q*), the participant had to decide whether the whole sequence seemed plausible (key p) or implausible (key q) (the ordering of the keys was counterbalanced between the participants). Both experiments were

Condition	P	Q
Logical (n=6)	*Jean a joué de la guitare* Jean played the guitar	*et Agnés a dansé le flamenco.* and Agnès danced flamenco.
Temporal (n=6)	*L'avion a atteri* The plane landed	*et les passagers sont descendus sur le tarmac.* and the passengers got off onto the runway.
Causal (n=6)	*Des pluies torrentielles se sont abattues sur le Jura* Heavy rainfall hit the Jura	*et l'électricité a été coupée.* and the electricity was cut off
Implausible (n=10)	*Les policiers ont attrapé le malfrat* The policemen caught the criminal	*et ont joué aux échecs avec lui.* and they played chess with him.

Table 6.2: Presentation of each type of sentence read

designed using the E-Prime 2.0 software (Schneider, Eschman & Zuccolotto 2002). All the sentences were presented in a random order, at the centre of a black computer screen in white text (Times New Roman font, size 18).

4.1.2 Results

The aim of this experiment was to try to isolate the mean reading time (RT) for each type of *et*, in order to compare them and test the predictions presented above. To do so, we measured the RT of the sequence composed of the conjunction *et* and the second sentence (*Q*) in the 'test' experiment, and of the second sentence alone (*Q*) in the 'control' experiment. In order to obtain a mean RT for each type of *et*, we subtracted the mean RT of each 'control' sentence (*Q*) from the RT of the corresponding 'test' sentences (*et Q*)—that is, $RT\ et = et\ Q - Q$. So, in the example *Marie a préparé les crêpes et Jean a passé l'aspirateur* (Marie made the pancakes and Jean did the vacuuming), we measured the RT of the second sentence with the conjunction (*et Jean a passé l'aspirateur*) and subtracted the mean RT of the second sentence without the conjunction (*Jean a passé l'aspirateur*) obtained by the participants from the 'control' experiment.

'Control' experiment An initial descriptive analysis allowed us to determine the mean reading time for each sentence (Q) in order to compute the *et* $Q - Q$ for each sentence: MLogical = 2374, SD = 1062; MTemporal = 2354, SD = 928; MCausal = 2016, SD = 796. In addition, a median test revealed no difference between the three conditions, $\chi^2 = 5.45, p = .065$.

'Test' experiment A Jonckheere-Terpstra test for ordered alternatives (Logical < Temporal < Causal) revealed a significant trend in the data, $T_{JT} = 111$, $z = 4.84, p < .001$ (see Table 6.3 and Figure 6.1). Associated post-hoc tests showed that the RT of (*et* $Q - Q$) of the logical condition was significantly smaller than the causal condition ($Z = 5.02, p < .001$), but not the temporal condition ($Z = .920, p = .54$). The same test revealed also a significant difference between the causal and temporal conditions ($Z = 3.97, p < .001$).

In addition, a Mann-Withney test showed a significant difference between symmetrical sentences (i.e. logical condition, $Mdn = -440$) and asymmetric sentences (i.e. causal and temporal conditions together, $Mdn = -284$), $Z = 3.36, p < .01$.

Condition	N	Mean	SD
Logical	267	-315.91	950.05
Temporal	258	-214.98	1024.46
Causal	230	123.04	1019.37

Table 6.3: Mean RT of (*et* $Q - Q$) as a function of the experimental condition (logical, temporal or causal).

4.2 Discussion

In short, the results obtained seem to indicate that symmetric propositions are processed faster than asymmetric ones. In particular, the logical interpretation of *et* is processed faster than the causal interpretation, and the temporal interpretation is also processed faster than the causal one.

The first observation is that the results of our experiment contrast with those obtained by Thompson et al. (2011, 2012). Since our study was carried out to verify reported findings in Thompson et al. (2011, 2012), and was thus designed to test a similar type of stimuli, we need to look more closely at the elements which could explain our results' differences.

Beyond the problem of stimuli, which we pointed out in section 2.3, the experiment conducted by Thompson et al. (2011, 2012) presents a major

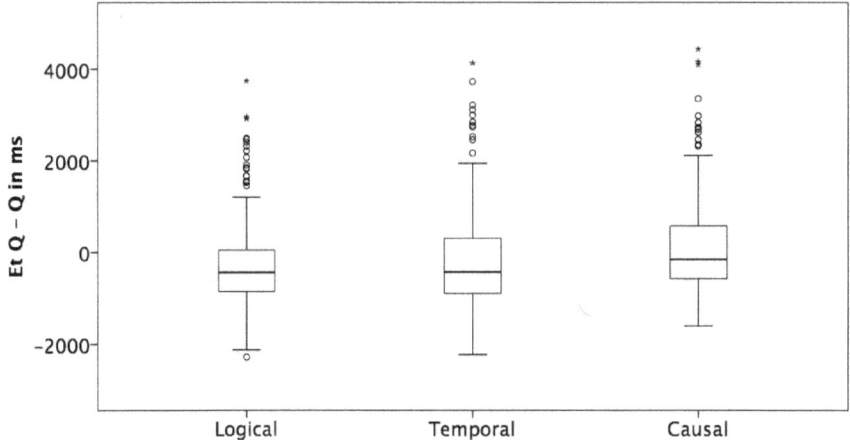

Figure 6.1: Box-plots of RT of *et Q – Q* according to the type of interpretation of the conjunction *et* (causal, logical and temporal). The black line indicates the median value. The boxes represent half of the sample, between the highest and lowest quartiles. The whiskers represent the most extreme values, excluding outliers (circles and asterisks).

methodological difference from ours, insofar as they used the *Rapid Serial Visual Presentation* paradigm and analysis of the production time, whereas we took the option of a *Self-Paced Reading* task. The former is more widely used in the research on lexical processing (Rayner & Sereno 1994), while the latter is more appropriate for studies on discourse comprehension (Garrod 2006). Aside from this, it should be pointed out that Thompson et al. (2011, 2012) do not provide statistical tests of their results. It is also important to emphasize that the small number of participants (eight) constitutes a weak point of their study.

The second observation is that the results obtained in the experiment on French partially support the predictions of pragmatic theories concerning the conjunction *and* (although see the discussion below). Logical interpretations (symmetric) were processed the most rapidly, followed by temporal and causal interpretations (asymmetric). Thus, our findings seem to be consistent with a series of experiments conducted on various logical terms presented in section 3.1 (see Noveck & Reboul 2008 for a summary). However, we did not observe a statistically significant difference between logical and temporal interpretations, despite pragmatic approaches' prediction that this difference should exist. So, why did we not find this? In the remainder of this section, we will try to explore possible reasons for this lack of difference in our results. The discussion will pinpoint some inherent difficulties in the definition of the logical interpretation of the conjunction *and* in natural language.

In contrast with the unambiguous vision we presented earlier, it turns out that logical and temporal interpretations are not so easy to tease apart in their natural language uses. In fact, the problem lies in the definition of the logical meaning of the conjunction *and* itself. As we explained in section 2, the logical meaning of conjunction is defined by its truth-table, from which it inherits the property of symmetry. Therefore, for a conjoined sentence to be classified as having a logical interpretation in practice, it is usually assumed that it must have the property of symmetry. However, our claim is that this criterion gives rise to a classification that is too coarse-grained. This is due to the fact that a speaker, when uttering a conjunctive statement, affirms in the vast majority of cases two things: i) that both conjoined propositions are true; and ii) that the denotations of these propositions (i.e. the corresponding states of affairs) are somehow related (some exceptions are dealt with below). This pragmatic use of conjunction takes into account two complementary aspects of the conjunction *and*: its logical function as a propositional operator; and its purely pragmatic ability to convey some relation between denotations of the conjuncts.

In its logical function, there is only one relation involved: the logical conjunctive relation[3], which has the property of symmetry. For instance, (20) is a typical example one might find in a logic textbook, but not really in everyday discourse.

(20)   Bern is a capital of Switzerland and 7 is a prime number.

The only two important points here are that both conjuncts are true, which makes the whole sentence true, and that it is possible to switch the order of conjuncts without any changes in the meaning. Besides this, there is no other relationship one could think of between the two conjuncts. This is therefore the logical relation of conjunction in its "pure" form, which uniquely determines the property of symmetry. We can also find more natural examples of the pure logical meaning of *and* in embedded clauses of the type discussed in section 2.2. The relevant example, with its syntactic analysis, is repeated below in (21).

(21)   a.   The newspaper reported that a new mayor was elected and that

---

[3] As we said in section 2, the conjunction *and* in logic is—strictly speaking—a binary operator which induces the logical relation of conjunction.

there was a riot.
b. ... reported [CP that ...] and [CP that ...]

In its pragmatic, everyday use, more relations are involved. Consider (22), where it is possible to reverse the conjuncts without changing the meaning of the whole *and*-sentence. The property of symmetry is conserved but, importantly, it is not uniquely determined by the logical relation of conjunction.

(22) John is playing basketball outside and Veronica is listening to music in her room.

What is crucial to note is that (22) conveys a temporal relation, that of simultaneity, which is also symmetric. In this case, the property of symmetry does not just pertain to the logical relation of conjunction, but is also attached to the real world relation of temporal simultaneity. Note that the same comment applies to example (1).

It is also important to observe that there are other types of relations which can be further conveyed with *and*, such as additive (23) or contrastive relations (24).

(23) John is tall and Veronica is tall.

(24) John is tall and Veronica is short.

Interestingly, there are languages which clearly distinguish between these two kinds of interpretations, using different conjunctions for them. In Polish, for instance, *i* serves—notwithstanding other nuances—to express additive relations (25), where *a* is employed for relations of contrast (26) (Wajszczuk 1984; see Sax 2014 for an analysis of Polish *a* in a Relevance Theory framework; see also Blakemore and Carston 1999, Blakemore 2002, Umbach 2004, Sawicki 2008 for different kinds of contrastive relations). Other languages, like Russian and Romanian, also make similar distinctions.

(25) a. Jan jest wysoki i Weronika jest wysoka.
      b. #Jan jest wysoki a Weronika jest wysoka.
      'John is tall and Veronica is tall.'

(26) a. Jan jest wysoki a Weronika jest niska.
      b. #Jan jest wysoki i Weronika jest niska.
      'John is tall and Veronica is short.'

It is worth asking what the relationship is between the purely logical relations which hold between the propositions and the relations between states of affairs that are denoted by these propositions, although we will not pursue such an investigation here. We will only observe that the property of symmetry, related to the logical conjunctive relation, may or may not coincide with the properties of relations conveyed by natural language interpretations of *and*. As it happens, some world relations are symmetric where others are not; thus, the properties of both logical and world relations are similar (i.e. symmetric) in some cases, while in others they are not. In the latter instance, the property of the given world relation prevails in the interpretation. It would be interesting to provide some formal answer to the question of why this is so. The answer to this question is beyond the scope of this contribution, and will thus be left for further investigation. Our attention will instead turn to a different question.

How can these observations explain our experimental results? Before answering this, we should note that the examples in our logical category conveyed a contrastive and/or simultaneous temporal relation, which is to say that they were not relation-free (other than a purely logical relation). If there is some conjunctive relation between the two conjuncts which is other than logical, we can suppose that a comprehender will grasp this relation in just the same manner as with temporal sequences of events or causal relations—that is, by inferring it. In this way, the comprehender enriches the meaning of the conjunction, just as in sequential temporal and causal interpretations. If this is correct, we should not be surprised that, despite the symmetry, we observe extra time in the processing of our logical condition.

In the light of this complementary explanation, the only certain conclusion we can draw from our experiment is that causal interpretations of *and* are processed at a slower rate than temporal (sequential and simultaneous) interpretations. Why is this so? In the remaining part of this discussion, we will propose an explanation according to the relevance-theoretic framework.

As we observed earlier, various relations that the utterances of *and*-sentences convey are the contributions to the explicit content of these utterances—that is, explicatures. The two relevant questions are the following: what is the difference between the two types of explicature; and what is the relationship between the two explicatures in causal *and*-conjoined sentences? As we said above, there are causal schemas which are invoked in causal interpretations of utterances of *and*-sentences (Carston 2002, Blochowiak 2014, 2016). There are also many temporal schemas which involve stereotypical sequences of events. Some of the temporal relations do not form stereotypical schemas, and the hearer can only recover a temporal relation on the basis of verbal tenses, aspectual classes and general world knowledge indicating how the two given eventualities can, a priori, follow one another or be parallel.

In the case of temporal explicatures, the procedure seems to be quite obvious: temporal relations appear in the presence of certain temporal schemas, if these exist, or merely on the basis of verbal tenses and aspectual classes in the absence of temporal schemas. Similarly, one could say that causal relations in *and*-conjoined sentences emerge in accordance with certain causal schemas. The crucial question is to know how temporal and causal explicatures interact or are related to one another in causal *and*-sentences.

Broadly speaking, we can think about two options: (i) the causal interpretation comes first, bypassing the temporal interpretation, which is to say that the hearer realizes right away that he is dealing with a causal interpretation of *and*; (ii) the temporal interpretation of *and* comes first, and the causal one at a second stage. The first option parallels the *causality-by-default* hypothesis (Sanders 2005), according to which comprehenders have a very strong cognitive drive to interpret—and often over-interpret—eventualities as causally related, which makes them very quick at detecting causality in the real world (Michotte 1963, Bechlivanidis & Lagnado 2016, Moors et al. 2017, *inter alia*), as well as at judging various descriptions of eventualities in language to refer to causal scenarios, independently of the presence of linguistic markers such as causal connectives (see, for instance, Sanders & Noordman 2000, Kuperberg et al. 2011, Mak & Sanders 2013). The second option echoes the *temporality-by-default* hypothesis proposed for *and* by Levinson (2000), which we described earlier: recall his argument that there is a cross-linguistically attested tendency to interpret two conjoined past-tense event descriptions as temporally successive by default (see also Lascarides & Asher 1993), and, if at all possible, as causally related.

From the processing point of view, the first option would predict shorter processing for causal interpretations of *and*, whereas the second option would predict shorter processing for temporal interpretations. Interestingly enough, our results seem to go against the *causality-by-default* hypothesis, as far as the causal interpretation of the conjunction *and* is concerned. This result is remarkable in itself, and will certainly need further experimental investigation. It seems that there is something in the meaning of *and* which prevents the causal interpretation from coming about immediately. Indeed, our findings seem to support the temporality-by-default hypothesis. A note of caution should nevertheless be sounded here: Levinson (2000: 123) talks specifically about temporal sequences of events described in the past tense, whereas in our experimental dataset we deal with temporal simultaneity (the logical condition) as well. In other words, in the case of Levinson's work, we should talk about the *temporal sequence* as the default interpretation of *and*. In that sense,

Levinsons's hypothesis finds partial confirmation in our results—that is, temporal sequential relations (temporal condition) are processed faster than causal relations (causal condition).

However, in order to provide a fuller explanation for our findings, we would like to propose an extension of the relevance-theoretic proposal concerning the conjunction *and*. Recall that temporal and causal interpretations of *and* are treated as explicatures in Relevance Theory, and arise in accordance with temporal or causal schemas whenever they are available, guaranteeing the requirement of optimal relevance. Our aim here is to specify how these two pragmatic inferences interact one with another. The prediction we formulated according to Carston's original proposal was that the temporal or causal interpretations can arise from the start, depending on the type of schema (temporal or causal respectively) available for a given *and*-conjoined sentence. However, for the causal interpretation of *and* to appear, we will argue below that the temporal inference—be it sequential or simultaneous—must have been the first to be drawn by the hearer. To illustrate this point, consider the following example, borrowed from Carston (2002: 236):

(27)   a.   She screamed and he hit her.
          b.   He hit her and she screamed.

In (27a), the interpretation is that he hit her as the causal result of her screaming, whereas in (27b), the opposite causal relation—that she screamed as a causal result of his hitting her—is recovered. This case is particularly interesting in the discussion of the possible interrelations between temporal and causal explicatures. Here, two possible causal schemas are available to us: (i) one can scream as a result of being hit by someone; or (ii) one can hit someone because he or she screamed. The crucial point we are making is that the choice of one causal schema over another is based on the temporal order of the events described in the conjuncts. In (27a), the hearer understands that first she screamed and then he hit her, and the presence of the relevant causal schema (in which the first event corresponds to the cause and the second to the causal consequence) ensures the causal interpretation in accordance with it. The opposite procedure applies in (27b), in which the hearer—according to the temporal order of the events, which is the converse of the previous order—selects the second causal schema. So, before concluding that temporal order plays a crucial role in establishing the causal relation in *and*-conjoined sentences, we need to ask the following question: what is the source of the inference of temporal sequence? The presence of the conjunction *and*, or maybe just past verbal tenses? An answer may come from the two following examples from

Carston (2002: 236), which present exactly the same sentences as (27), but without the conjunction *and*.

(28) a. She screamed. He hit her.
      b. He hit her. She screamed.

As Carston notes, the causal interpretations are not as clear-cut as in (27). As such, in the presence of conflicting causal schemas, the simple alignment of events in juxtaposed sentences is not enough to impose the temporal sequencing of events. Therefore, we can conclude that it is indeed the conjunction *and* which 'forces' the temporal order of events described in the conjuncts.

What are the implications for our initial question about the interaction between the two explicatures—the temporal and the causal—in causal interpretations of *and*-sentences? Our proposal is that the temporal explicature in the case of *and*-conjoined sentences is derived first, and this stage is necessary for the drawing of the causal explicature. In other words, a temporal inference in the form of an explicature is the starting point, and a plausible causal explicature is calculated on the basis of the temporal one, given that a relevant causal schema is available to the hearer. From the processing point of view, this proposal predicts that temporal interpretations of *and* will take less time, as there is only one explicature for the hearer to recover, while causal interpretations of *and* will be costlier, since the hearer needs to draw two pragmatic inferences (a temporal explicature and a causal explicature).

What about simultaneous cases? In our experimental dataset, the stimuli from the logical condition refer for the most part to simultaneous temporal relations, or at least to temporal inclusion, and, as we saw, do not differ statistically from the temporal sequences (temporal condition). Thus, it seems that the two types of temporal relations behave similarly as far as the cognitive processing of *and*-conjoined sentences is concerned, which would be consistent with our hypothesis that any type of temporal relation has to be recovered by a pragmatic inference, which is a cognitively costly process.

And what of the causal relations based on simultaneous temporal relations? Even in the case of simultaneity, it seems that an order has to be recovered for the causal relation to arise. For instance, the causal relation is present in (29a) but absent in (29b), where the two conjuncts are flipped around.

(29) a. The atmospheric pressure is low and I have a headache.
      b. I have a headache and the atmospheric pressure is low.

This example suggests that, even in the case of causality with states, the order in which each state began must be recovered to arrive at a causal interpretation—that is, the state in which the atmospheric pressure is low (or the event in which it dropped) started before the state in which I started to have a headache, even if they continue in parallel for some period of time and support one another causally (cf. Blochowiak 2009 for more detail). While we did not verify causal relations based on simultaneous temporal relations in the experiment presented here, the prediction would be similar to causal relations based on temporal sequences. In both types of situation, the hearer has first to establish the temporal coordinates of the eventualities described in the conjuncts, and she can search for a causal link afterwards.

4.3 Conclusions

This paper presented a self-paced reading experiment which aimed to verify experimentally the predictions of syntactic and pragmatic theories concerning the conjunction *and*. On the one hand, according to a syntactic proposal (Bjorkman 2010), there is a structural difference between the symmetric uses (including logical interpretations) implying bigger syntactic structures (CP) and asymmetric uses (including temporal and causal interpretations) composed of smaller syntactic structures (TP). Thus, the syntactic approach predicts shorter processing for asymmetric interpretations (temporal and causal) and longer processing for symmetric ones (logical). On the other hand, pragmatic theories claim that the meaning of the conjunction *and* in natural language is based on the meaning of the logical conjunctive operator ∧, which is symmetric. Hearers pragmatically infer other types of interpretation, such as temporal or causal ones, by general principles of pragmatics. In accordance with this general pragmatic comprehension procedure, a relevance-theoretic account predicts that semantic/logical interpretations should be processed faster than pragmatic ones (temporal and causal). A different position is held by Levinson (2000), for whom pragmatic interpretations come first by default, and in the particular case of *and*, the temporal relation is claimed to be the first recovered by the hearer.

The results of the experiment conducted by Thompson et al. (2011, 2012) on the English conjunction *and* point towards a partial confirmation of the syntactic theory. However, these results do not seem to be easily generalizable, given the problems regarding the construction of the stimuli and the small number of participants in the sample size (cf. Section 3.2).

The results of the experiment we conducted on the French conjunction *et* were the opposite of those obtained for English by Thompson et al. (2011, 2012). Logical and temporal interpretations were processed the fastest, followed by causal interpretations, with a statistically significant difference between the logical and causal interpretations but no statistically significant difference observed between logical and temporal conditions.

The fact that we could not find a statistically significant difference between logical and temporal *and* is interesting, as it highlights a theoretical problem of the definition of the logical interpretations of the conjunction *and* in natural language. As it turns out, the property of symmetry—usually taken as the criterion for classifying sentences with *and* as logical—is problematic, as some relations which *and* can convey in natural language (such as temporal simultaneity or inclusion) have the property of symmetry independently of the property of symmetry coming from logic. In other words, the stimuli from our logical condition could in fact be interpreted as instances of temporal simultaneous relations.

In the light of this finer-grained analysis of the data from the logical condition, we can reinterpret our findings as follows: we observed a statistically significant difference between causal interpretations on the one hand and temporal interpretations (sequential and simultaneous) on the other. How does this new restatement fit the predictions? First, we should observe that neither the syntactic account nor the relevance theoretic approach differentiates between the temporal and causal types of interpretation. The only prediction to have been partially verified is Levinson's 2000 proposal, which put forward the temporality-as-default hypothesis. However, Levinson talked about the default interpretation for two conjoined past-tense event descriptions as temporally successive. Yet, in our experimental dataset, we also have simultaneous or overlapping temporal relations (i.e. logical condition) which seem to pattern with sequential ones. This observation has led us to formulate a new hypothesis, which is a refinement of the relevance theoretical account for conjunction. Nevertheless, it should be stressed that our hypothesis is tentative. It is open to further empirical verification, and holds only for as long as the results we obtained can be confirmed.

We followed a relevance-based analysis of *and*, which argues that its temporal and causal interpretations arise as explicatures—that is, they are pragmatically inferred contributions to the explicit content of utterances of *and*-sentences. Our proposal is that these explicatures are not independent as far as the causal interpretations of *and* are concerned. In particular, for a hearer to derive a causal interpretation, he has first to infer the temporal ordering of the events described in the conjuncts. Consequently, in the case of temporal interpretations of *and*, only one explicature is to be drawn by the hearer (i.e. the temporal explicature, be it sequential or simultaneous), whereas in the case of the causal interpretation of *and*, two explicatures are computed (i.e. the temporal explicature augmented by the causal explicature), provided that the relevant causal schema is accessible to the hearer. Our hypothesis is based on the observation that when two causal schemas are available (the event described in the first conjunct could have caused the event described in the second conjunct or *vice versa*), it is the temporal order of eventualities which imposes the choice of one schema over another (cf. example (27) and subsequent discussion). From the point of view of processing, our proposal's prediction is clear. Temporal interpretations of *and* should be processed faster than causal interpretations, as the hearer has to infer one explicature when only temporality is involved, where in the case of causal *and*, the hearer has to recover the causal explicature according to the temporal explicature. It is important to note another prediction which follows from our proposal: since it is the conjunction *and* which 'forces' the initial temporal interpretation, in juxtaposed sentences (cf. example (28)), we do not expect the temporal interpretation to arise before the causal interpretation. In summary, we predict that causal interpretations will be processed faster than temporal interpretations in the case of juxtaposed sentences. We will leave this prediction for our future investigations.

References

Bechlivanidis C. & D. Lagnado (2016). "Time reordered: Causal perception guides the interpretation of temporal order". *Cognition* 146: 58–66.

Bjorkman B.M. (2010). "A syntactic correlate of semantic asymmetries in clausal coordination". *Proceedings of NELS* 41.

Blakemore D. & R. Carston (1999). "The pragmatics of *and*-conjunctions: The non-narrative cases". *UCL Working Papers in Linguistics* 11: 1–20.

— (2005). "The pragmatics of sentential coordination with *and*". *Lingua* 115(4): 569–589.

Blochowiak J., T. Castelain & J. Moeschler (2015), "Les interprétations logiques, temporelles et causales de la conjonction. Une perspective expérimentale." *Nouveaux cahiers de linguistique française* 32: 71–83.

Blochowiak J. (2009). "La relation causale, ses relata et la négation". *Nouveaux cahiers de linguistique française* 29: 153–175.

— (2014). *A theoretical approach to the quest for understanding. Semantics and pragmatics of whys and becauses*. PhD thesis, Department of Linguistics, University of Geneva.

— (2016). "A presuppositional account of causal and temporal interpretations of *and*". *Topoi* 35: 93–107.

Bott L. & I. Noveck. (2004). "Some utterances are underinformative: The onset and time course of scalar inferences". *Journal of Memory and Language* 51(3): 437–457.

Breheny R., N. Katsos & J. Williams. (2006). "Are generalised scalar implicatures generated by default? An on-line investigation into the role of context in generating pragmatic inferences". *Cognition* 100(3): 434–463.

Carston R. (1988). "Implicature, explicature, and truth-theoretic semantics". In R.M. Kempson (ed). *Mental representations: the interface between language and reality.* Cambridge: Cambridge University Press. 155–181.

— (1993). "Conjunction, explanation and relevance". *Lingua* 90(1-2): 27–49.

— (2002). *Utterances and Thoughts. The Pragmatics of Explicit Communication.* Oxford: Blackwell.

Forster K. (1970). "Visual perception of rapidly presented word sequences of varying complexity". *Perception & Psychophysics* 8(4): 215–221.

Garrod S. (2006). "Psycholinguistic research methods". In M. Traxler & M.A. Gernsbacher (eds), *Handbook of psycholinguistics.* San Diego: Academic Press. 455–503.

Grice H. P. (1975). "Logic and conversation". In P. Cole & J.L. Morgan (eds) *Syntax and semantics, 3: Speech acts.* Amsterdam: Elsevier. 41–58.

Grice H. P. (1989). *Studies in the Way of Words.* Cambridge MA: Harvard University Press.

Grisot C. & Blochowiak J. (under revision). Temporal connectives and verbal tenses as processing instructions: evidence from French. *Cognitive Pragmatics.* De Gruyter.

Grisot C. (2015). *Temporal reference: empirical and theoretical perspectives. Converging evidence from English and Romance.*, PhD thesis, Department of Linguistics, University of Geneva.

Guasti M.T., G. Chierchia, S. Crain, F. Foppolo, A. Gualmi & L. Meroni (2005). "Why children and adults sometimes (but not always) compute implicatures". *Language and cognitive Processes* 20(5): 667–696.

Horn L. (1972). *On the semantic properties of the logical operators in English.* Doctoral dissertation, University of California, Los Angeles.

Kuperberg G.R., M. Paczynski, & T. Ditman (2011). "Establishing causal coherence across sentences: An ERP study". *Journal of Cognitive Neuroscience* 23: 1230–1246.

Lascarides A. & N. Asher (1993). "Temporal interpretation, discourse relations and commonsense entailment". *Linguistics and philosophy*, 16(5): 437–493.

Levinson S.C. (1983). *Pragmatics.* Cambridge: Cambridge University Press.

— (2000). *Presumptive meanings: The theory of generalized conversational*

implicature. Cambridge, MA: MIT press.
Mak W.M. & T.J.M. Sanders (2013). "The role of causality in discourse processing: Effects of expectation and coherence relations". *Language and Cognitive Processes* 28(9): 1414–1437.
Michotte A. (1963). *The Perception of Causality.* Andover: Methuen.
Moeschler J. (2000). "Le Modèle des Inférences Directionnelles". *Cahiers de linguistique française* 22: 57–100.
— (2002). "Economy and Pragmatic Optimality: the Case of Directional Inferences". *Generative Grammar in Geneva* 3: 1–20.
— (2010). "Negation, scope and the descriptive/metalinguistic distinction". *Generative Grammar in Geneva* 6: 29–48.
Moors P., J. Wagemans, & L. de Wit (2017). "Causal events enter awareness faster than non-causal events". *PeerJ* 5:e2932.
Nieuwland M., T. Ditman & G. Kuperberg. (2010). "On the incrementality of pragmatic processing: An ERP investigation of informativeness and pragmatic abilities". *Journal of Memory and Language* 63(3): 324–346.
Noveck I. (2001). "When children are more logical than adults: Experimental investigations of scalar implicature". *Cognition* 78(2): 165–188.
Noveck I. & Reboul A. (2008). "Experimental pragmatics: A Gricean turn in the study of language". *Trends in Cognitive Sciences* 12(11): 425–431.
Noveck I., C. Chevallier, F. Chevaux, J. Musolino & L. Bott. (2009). "Children's enrichments of conjunctive sentences in context". In P. De Brabanter & M. Khissine. *Utterance Interpretation and Cognitive Models. Current Research in the Semantics/pragmatics Interface* (20). Bingley: Emerald. 211–234.
Papafragou A. & J. Musolino (2003). "Scalar implicatures: experiments at the semantics–pragmatics interface", *Cognition* 86(3): 253–282.
Pouscoulous N., I. Noveck, G. Politzer & A. Bastide (2007). "A developmental investigation of processing costs in implicature production". *Language acquisition* 14(4): 347–375.
Posner R. (1980). "Semantics and pragmatics of sentence connectives in natural language". In J. Searle, F. Kiefer & M. Bierwisch (eds). *Speech act theory and pragmatics.* Dordrecht: Reidel. 169–203.
Rayner K. & S.C. Sereno (1994). "Eye movements in reading". In M.A. Gernsbacher (ed). *Handbook of psycholinguistics.* San Diego: Academic Press. 57–81.
Sanders T.J.M. (2005). "Coherence, causality and cognitive complexity in discourse". In *Proceedings of the First International Symposium on the exploration and modelling of meaning* (SEM-05). 105–114.

Sanders T.J.M. & L.G.M. Noordman (2000). "The role of coherence relations and their linguistic markers in text processing". *Discourse Processes* 29: 37–60.
Saussure L. de (2003). *Temps et pertinence: éléments de pragmatique cognitive du temps*. Bruxelles: De Boeck Duculot.
Sawicki L. (2008). *Toward a Narrative Grammar of Polish*. Warszawa: Warsaw University Press.
Sax D. (2014). "Osobliwy przypadek polskiego spójnika *a*: analiza z perspektywy teorii relewancji Sperbera i Wilson", Talk given at Instytut Jzyka Polskiego, Warsaw University on May 22, 2014.
Schmerling S. (1975). "Asymmetric conjunction and rules of conversation". *Syntax and semantics* 3: 211–231.
Schneider W., A. Eschman & A. Zuccolotto (2002). *E-Prime user's guide*. Pittsburgh: Psychology Software Tools Inc.
Sperber D. & D. Wilson (1986/1995). *Relevance: Communication and Cognition*. Oxford: Blackwell/Cambridge, MA: Harvard University Press. 2nd edition.
Su Y. E. (2014). "The Acquisition of Logical Connectives in Child Mandarin". *Language Acquisition* 21(2): 119–155.
Thompson E., J. Collado, M. Omana, A. Yousuf-Litle (2011). "The processing of asymmetric and symmetric sentential conjunction". *Proceedings of the 4th ISCA workshop ExLing 2011*, Paris, France. 131–134.
— (2012). "The processing of asymmetric and symmetric sentential conjunction." *International Journal of Language Studies* 6(4): 25–40.
Tiel B., & Schaeken W. (2017). "Processing conversational implicatures: alternatives and counterfactual reasoning." *Cognitive science* 41(S5): 1119–1154.
Umbach C. (2004). "On the Notion of Contrast in Information Structure and Discourse Structure". *Journal of Semantics* 21(2): 1–21.
Wajszczuk J. (1984). "Metatekstowe *szwy* tesktu. Casus: polski spójnik *a*". In L. Lönngren (ed). *Polish Text Linguistics: the third Polish-Swedish Conference held at the University of Uppsala, 30 May4 June 1983*. Uppsala: Slaviska institutionen vid Uppsala universitet. 53–75.
Wilson D. & D. Sperber (1993). "Linguistic form and relevance". *Lingua* 90: 1–25.

Chapter 7
Contexts, Biases, and Reflexivity*

Eros Corazza

Abstract In his book *Meaning without Truth* (2013), Predelli offers a unifying theory of biases dealing with seemingly various phenomena often relegated to the pragmatics dustbin (nicknames, slurs, answering machines/post-its, vocatives, etc.). In this paper I briefly explain Predelli's theory and, focusing on slurs, I argue that Predelli's theory of bias cannot be explained away using Grice's notion of conventional implicatures, but may be explained in terms of Gricean generalized conversational implicatures. I propose an utterance-bound pluri-propositionalist outline (inspired by the work John Perry and Kepa Korta) of a theory of communication and show it can handle various biases.

Keywords Theory of biases · Slurs · Generalized conversational implicatures · Pluri-propositionalism ·

---

*A short version of this paper was presented at the 2014 ALFA conference in Fortaleza, Brazil. Comments and discussions with Kepa Korta, John Perry, Stefano Predelli, Marco Ruffino, Pierre Saint-Germier, Ludovic Soutif, Richard Vallée and Tim Williamson have been helpful. The research for this paper has been partially sponsored by a Social Sciences and Humanities Research Council of Canada grant, SSHRC (Standard Research Grant, 410-2010-434), the Spanish Ministry of Science and Innovation (FFI2009-08574), and the Basque Government (IT323-10).

1 Introduction

In a recent, welcome contribution, Predelli (2013) presents a fruitful and powerful framework on how truth-conditional semantics can deal with non-truth-conditional meaning. One of the gifts of Predelli's contribution is that it presents a unifying theory dealing with seemingly various phenomena often relegated to the pragmatics dustbin. Thus, nicknames, slurs, answering machines/post-its, vocatives, etc. get captured within the semantic framework, the theory of biases, presented.

In this paper I will briefly explain Predelli's theory in section 2. In section 3, focusing on slurs, I will argue how Predelli's theory of biases cannot be explained away using Grice's notion of conventional implicatures. In section 4, I will show how Predelli's theory of biases may be explained in appealing to Grice's generalized conversational implicatures. Finally, in section 5, I will propose an utterance-bound pluri-propositionalist outline (inspired by Perry 2002/12 and Korta & Perry 2011) of a theory of communication. I will show how such a theory can handle various biases in quite a straightforward, possibly cognitively more plausible, way.

It may turn out that the proposal I will sketch is but a notational variant of Predelli's theory of biases. If so, then Predelli's theory is a valuable addition to the pluri-propositionalist theory of communication, or so I will try to argue.

2 Predelli's Theory of Biases

Predelli's main strategy is that we can deal with what is traditionally characterized as non truth-conditional meaning by putting some constraints on use. That is, a linguistic expression, on top of coming equipped with a Kaplanian character (see Kaplan 1977), comes equipped with what Predelli characterizes as bias.[1] The utterance of a sentence like

(1) I am busy.

said by John, expresses the (singular) proposition that John is busy in virtue of the character of the first person pronoun 'I' selecting the speaker as its value. The very same sentence, uttered by Jane, expresses the proposition that Jane is

---

[1] Cf. Kaplan's (1977) content/character distinction. The character (or linguistic meaning) of an indexical can be represented as a function taking as its argument the context and delivering as its value the content or referent. Thus, the character of 'I' can be represented as 'the agent of this utterance', the character of 'you' as 'the addressee of this utterance', the character of 'today' as 'the day of this utterance', etc.

busy. On the Kaplanian account, the character of 'I' is represented by a function that takes as argument the *agent*, and gives as value the referent (agent). Were this sentence written on a post-it (or registered on an answering machine), though, 'I' would not necessarily select the writer and/or speaker as its value—one may well use a pre-recorded answering machine playing "I am busy" or borrow a post-it written by someone else. In such a scenario, 'I' would not select the writer or speaker as its agent.[2] This helps us to stress how the notion of *use* needs to enter the scene. For, a sentence like (1) when used in a given context may not pick up its producer. In a face-to-face linguistic exchange, or face-to-face context of use, (1) picks out the speaker as the agent and, thus, as the referent entering the proposition expressed (roughly, what is said). In a post-it situation, or text-message context of use, for instance, (1) need not pick up the producer (e.g. the writer) as the agent of it. The changes in interpretations do not rest on a change in the character of the expressions used ('I' is nether ambiguous nor a case of polysemy), but on a change in the relevant feature of the context of use. Predelli's theory of biases attempts to handle cases like this, for in the context of a will (or a letter) the signature is considered as a device of obstinacy. That is to say, the audience is removed from the task of identifying the pertinent contextual aspect fixing the reference of the indexical expression. The contextual aspect (e.g. the agent, time or location) is settled by the context of use, so that the indexical expression can only be correctly used to refer to a particular individual time or location that is fixed in the context of use.[3]

The general moral seems to be that a plausible theory of communication needs a theory of use. For, it is the latter that helps us to distinguish how the production of a given sentence can mean different things according to the type of use it is embedded into. Following Predelli (2013: 9-7), an expression *e* and a context *c* can be characterized as a use only if *c* is a context of use (there are contexts, e.g. *silent* contexts, that are token-free and, as such, cannot count as context of use). Furthermore, a token presupposes the existence of *intentional* agents—the movement of an ant on a dusty desk giving the impression "I am busy" is not classified as a token.[4] To summarize:

---

[2] The literature on these phenomena is now rich. See, for instance, Sidelle (1991), Smith Q. (1989), Vision (1985), Predelli (1998), Corazza et al. (2001), among others.

[3] A different way to handle these kinds of problems (what Predelli characterizes as obstinate uses) would be to argue that in such scenarios the indexical expression comes close to working as an anaphoric pronoun coindexed (and, thus, coreferential) with the term (e.g. date, signature) settled by the context of use. See Corazza (2002, 2004).

[4] As Putnam once claimed, an ant impressing with her movement on the sand what we take to be a picture of Churchill does not count as a picture of Churchill for the ant did not intend it

- A context *c* belongs to a context of use *CU* only if there exist (have existed, or will exist) tokens of expressions in *c*

and

- A context *c* belongs to a context of use *CU* only if there exist (have existed, or will exists) intentional agents in *c*.

To put it into a nutshell, for any expression, there is a class of contexts of (appropriate) use of that expression.

Consider utterances of sentence like:

(2)   I exist.

(3)   I am here now.

Every time one utters them one makes a trivially true statement. Yet one is not necessarily a being and an utterance of (2) is not necessarily a truth. For instance, one may begin one's will by writing (or recording) "I do not exist" or "I am not here now". When read or heard by the heirs, they are true. Their truth is not determined by a change in the character of the indexicals 'I', 'now' and 'here'; whether uttered in a face-to-face context or written and/or recorded in a will, the character of 'I' and 'now', for instance, is the same. That is, (2) and (3) are not true in virtue of their character alone. Their truth (or falsity) depends on the context of use. In the face-to-face context of use they are self-verifying. If we switch the context of use, e.g. the text-message context (as it would be if they were uttered or recorded in a will) they would be false, not because the character of 'I' or 'now' changes, but because of the change in the context of use.

One of Predelli's main points is that to deal with non-truth-conditional meaning we must take on board constraints on appropriate contexts of use. For, on top of coming with its truth-conditional profile (roughly, a Kaplanian character), expressions are encrypted with what Predelli characterizes as a *bias*, i.e. a condition in which a context counts as an appropriate context of use for that expression. The *conventional* meaning of an expression can thus be considered as a pair comprising the character of the given expression *and* the bias it conveys:

> [I]t is part and parcel of the meaning of 'hurray' that it may appropriately be used only by speakers favorably disposed toward a

---

to be interpreted this way.

certain event. As a result, the class of contexts of use for 'hurray' may contain only contexts with at least occasionally elated agents, not for reasons deriving from the properties of the type of use in question, but as a result of features encoded within the conventional profile of that interjection. (Predelli 2013: 62)

Thus, a context c belongs to the context of use of 'hurray', CU(hurray), only if the agent of c is favorably disposed toward something or other. In the case of 'hurray', like 'alas', the expression has a null character, yet it constrains the context of use for it settles the speaker's (agent's) attitude *vis-à-vis* the event expressed by the utterance embedded within it. To borrow from Frege (1892) we can say that it pertains to the coloration (or tone) of the relevant utterance.[5] That is to say, the coloration conveyed by an expression directs us toward the context of use. The same, or similar enough, story can be told about expressions such as 'tummy', 'abdomen', 'stomach', and 'belly'. Though coextensive, they differ in their coloration. If one uses 'tummy' one indicates a child-directed speech, while if one uses 'abdomen' one indicates a professional (e.g. medical)-directed speech. Within the framework proposed by Predelli (2013: 84 ff.) these terms come out to be biased inasmuch as they indicate the appropriate context of use. Roughly, a context c belongs to the context of use for 'tummy' if the addressee in that context is a child. This is, Predelli claims, a case of non-denotational bias.

In summary, while the character of an expression can be represented (along the lines of Kaplan's theory of demonstratives) by a conventionally assigned function that determines the content entering the proposition expressed, the bias of an expression is a constraint conventionally assigned to a term that fixes its relationship to the context of use of that expression, be it a token word or an utterance.[6] The picture emerging is that expressions have both a bias and a character: though some expressions, like 'I', 'now', 'house' or 'water', have a null bias, others, like the interjections 'hurray' or 'alas', have a null character.

Predelli's theory of bias provides the framework upon which a general theory of communication can be proposed. Another example worth mentioning is

---

[5] Speaking about translation, Frege claims that "the possible differences here belong also to the coloring and shading which poetic eloquence seek to give to the sense. Such coloring and shading are not objective, and must be evoked by each hearer or reader according to the hints of the poet or the speaker" (Frege 1892: 61). Very roughly, like Frege's notion of coloration Predelli's theory of bias must be dealt with at the non-truth-conditional level. Yet, unlike Frege's notion, the truth value of the proposition semantically expressed depends on aspects pertaining to coloration, for the latter enters the context of use.

[6] While the bias introduced by 'alas' operates on the whole utterance, the one introduced by a slur (e.g. 'fag', 'wop', 'boche', etc.) operates on the word/slur.

the use of derogatory language. If we consider, for instance, slurs like 'wop' or 'kraut' and their correspondent neutral (bias-silent) 'Italian' and 'German', we can say that 'wop' and 'kraut' determine a context of use where the speaker expresses a negative attitude regarding the class of individuals singled out by the word, in our example Italians and Germans, respectively. To illustrate this let us consider:

(4)   Hitler was a German Nazi.

(5)   Hitler was a kraut Nazi.

With (5), independently of the fact that Hitler was, without doubt, a despicable person, one also manifests one's xenophobic attitude *vis-à-vis* Germans in general (and not only Hitler), that would not be manifested if one were to utter (4) instead.[7] A German audience need not be a neo-Nazi, let alone a sympathizer of Hitler's crimes, to feel offended on hearing (5). Similarly if an Italian hears "Fritz is not a wop", she is likely to feel offended even if the slur is embedded within a negation. On the other hand, our Italian audience would not feel offended in hearing: "Fritz is not Italian". This seems to suggest, at least at the intuitive level, that there is a difference in meaning between the slur 'wop' and the neutral 'Italian'. The question that springs to mind is how to account for this difference. Predelli's theory of bias attempts to capture this intuition. Since the *bias* is *conventionally* conveyed, i.e. associated with the term, it follows that utterances of sentences like (5) conventionally convey the negative (racist and xenophobic) attitude—the negative bias—of the speaker. Yet, at least at first sight, one can utter:

(6)   Fritz is German, he is *not* a kraut.

without contradicting oneself.[8] What the speaker would express is his/her unfavorable attitude toward the use of the slur 'kraut'. In replying that Fritz is not a kraut, one is not denying the fact that Fritz is German. What one objects to is the use of the slur 'kraut'. In other words, for this to be considered an acceptable and appropriate utterance in the mouth of a non-racist speaker, it ought be understood in an echoic way, as a refutation of the use of the slur-word. If so, it should be cashed out as a case of metalinguistic negation, and treated in the

---

[7]Following Predelli the *bias* is conventionally conveyed, i.e. associated with the term. Yet it is not determined by the character of the term. 'German' and 'kraut' have the same character. What differentiates them is the context of use. While the former is bias-silent, the slur points toward a context of use suggesting that the utterer has a negative bias regarding Germans.

[8]In uttering a sentence like this, the slur-dismissing speaker is likely to put the voice stress on the negation and, in so doing, to trigger a meta-linguistic interpretation.

way we would treat "Jane is not a doctor, she is a cardiologist" meaning that Jane is better characterized as a cardiologist than as a doctor or "Jane is not intelligent, she is a genius" meaning that Jane is better characterized as being a genius rather than merely intelligent.

3 Biases *qua* Conventional Implicatures

If we focus on the negative connotation and/or attitude triggered by slurs, a natural way to deal with them (and possibly with all of what Predelli characterizes as biases) would be to adopt Grice's (1967/87) distinction between conventional implicatures and conversational implicatures. One could claim (e.g. Potts 2007 and Williamson 2009, among others) that slurs trigger conventional implicatures. Although 'and' and 'but' are truth-conditionally equivalent (and can be represented by '&'), utterances of sentence containing 'but', unlike sentences containing 'and', conventionally trigger that there is a contrast between the two conjuncts. In hearing "John is rich but he is honest", on top of computing that John is rich and that he is honest, one computes that there is a contrast between being rich and being honest, i.e. that rich people are not usually honest.

If the negative attitude conveyed by a speaker uttering a slur is characterized as a conventional implicature, in hearing (4) one would thus compute that Hitler was a German Nazi and conventionally implicate that the speaker of it has a negative attitude toward Germans. This seems, no doubt, a promising strategy: while 'German' and 'kraut' are coextensive, utterances containing the slur conventionally implicate that the speaker has a negative attitude toward Germans. Like a conventional implicature triggered by 'but', the one triggered by 'kraut' is not cancellable, inasmuch as it is triggered by the meaning of the term used. One cannot successfully deny one's negative attitude toward Germans by saying, e.g.: "Fritz is a kraut but Germans are my favorite/preferred people". Furthermore, like a conventional implicature, the implicature triggered by 'kraut' is not calculable. Besides, *qua* conventional implicature, it is neither context-dependent, nor dependent on the intention of the speaker.

There is, though, a difference between a slur and a device for conventional implicatures, say, 'but'. Following Grice, a conventional implicature is encoded, *viz.* it is triggered by the conventional meaning that the utterance inherits from the sentence (the type) uttered. At first sight, this seems to support the view that the implicature triggered by a slur is a conventional one insofar as the slur-word *conventionally* encodes the information that it is a contemptuous word. Yet, a conventional implicature, unlike a conversational one, cannot be cancelled (or defeated). One cannot say "John is rich but he is honest and all rich people are honest" without contradicting oneself.[9] Furthermore, while a competent speaker using or hearing a sentence containing 'but' *knows* that there is a contrast between the first and the second phrase, a competent speaker may not know exactly what the negative meaning conveyed by slurs like 'wop' or 'kraut' is. One who knows that 'wop' is a negative term to characterize Italians may not be classified as incompetent with English if she does not know that it derives from 'without papers', let alone that the targeted individuals characterized as 'wop' are associated with the negative stereotypes of being lazy and having a relaxed attitude toward the moral dictates. In other words, one may be competent with 'wop' being a slur without knowing all the stereotypes it may convey. If one is asked why 'wop' is a negative term for Italians, one may not be aware of all the negative features or stereotypes concerning Italians it may carry.[10]

Another question, more important for the purpose of this paper, that comes to mind is whether a theory treating slurs as cases of conventional implicatures generalizes so as to account for the many other cases of biases presented by Predelli. As I stated, one of the main gifts of Predelli's theory of biases is that it is a *unified* theory. Could, for instance, the positive bias introduced by 'hurray' (or the negative one introduced by 'alas') be treated as a conventional implicature? Can the analysis of:

(7)   Hurray, John has been promoted.

(with 'hurray' having, as Predelli claims, a null character) be cashed out in the way Grice analyzes 'but'? That is, does the speaker say (truth-conditionally convey) that John has been promoted and conventionally implicate that s/he has a positive attitude toward John being promoted?

Furthermore, if one (like Bach 1999, cf. also Corazza 2012) argues that conventional implicatures are a myth, one is tempted to resist the treatment of

---

[9] One could argue that conventional implicatures can be cancelled as could be the case in the following discussion-fragment: A: "All rich people are dishonest"; B: "No, John is rich but honest. In fact, all rich people are honest". If this is the case, though, alleged conventional implicatures turn out to be *conversational* implicatures. If so, then the treatment of slurs as generalized conversational implicatures that I will defend turns out to be the right one.

[10] As Vallée suggests, in discussing Williamson's (2009) account: "In learning a slur, however, we do not learn a very specific negative feature conventionally implicated by that word. Moreover, hearing a slur, we do not try to identify the single negative feature conventionally implicated by that term. It is unclear to me, and it is not backed by any arguments in the literature, whether 'boche', or any slur for that matter, semantically conveys a well-defined, specifiable content as is the case with what 'but' conventionally implicates" (Vallée 2014: 88).

slurs as conventional implicatures.[11] Bach claims that a speaker of:

(8)   John is rich but he is honest.

ends up saying two things (expressing two propositions), i.e.:

(9)   a.   That John is rich and that John is honest.
       b.   That being rich contrasts with being honest.

According to Bach, there is a simple test we can apply to decide whether something belongs either to what is said or to some other pragmatic feature, i.e. the *Indirect Quotation* (IQ) test.

> (IQ)   An element of a sentence contributes to what is said in an utterance of that sentence if and only if there can be an accurate and complete indirect quotation of the utterance (in the same language) which includes that element, or a corresponding element, in the 'that'-clause that specifies what is said. (Bach 1999: 340)

The IQ test "only excludes elements that do not contribute to what is said in the sense of propositional content" (Bach 1999: 340). Since alleged conventional implicature devices, ACIDs, contribute to what is said, they are not devices of conventional implicature.[12] To illustrate this, Bach invites us to consider:

(10)   a.   Jane: "Shaq is huge *but* he is agile".
       b.   Jane said that Shaq is huge but that he is agile.[13]

---

[11] Carston (2004) also express skepticism regarding conventional implicatures.

[12] For a defense of the IQ test see also Cappelen & Lepore (2005). If one, following Perry (1986; see also Korta & Perry 2011) accepts the presence of unarticulated constituents into the proposition expressed (the Gricean what is said) one would likely reject the IQ test and possibly argue that conversational implicatures would contribute into the proposition expressed, i.e. that the minimal proposition gets enriched *via* alleged conversational implicatures. Although this move may seem, at first sight, plausible, I do not see how it could deal with biases in general. In saying, e.g. "Fritz is a kraut", would one express the proposition that Fritz is German and that he has a negative (xenophobic) attitude toward Germans? This runs against our intuitions for, strictly speaking, all one says is that Fritz is German; one does not also say that he dislikes Germans. In other words, can an alleged conventional implicature enter, as an unarticulated constituent, the proposition expressed? This debate, as interesting as it may be, transcends the scope of the present chapter. As I will argue, biases can be explained in appealing to reflexive contents without having them enter the official content.

[13] It is worth noticing that for the IQ test to work, 'but' must be understood as appearing

(11) a. Jane: "*Even* Shaq can make some free throws".
b. Jane said that *even* Shaq can make some free throws.

In (10) the contrastive part of 'but' contributes to what is said. The alleged conventional implicature belongs, *de facto*, to what has been said by Jane. In (10a) Jane actually said two things:

(12) a. That Shaq is huge and that Shaq is agile.
b. That being huge contrasts with being agile.[14]

If Bach is, as I believe, right concerning (alleged) conventional implicatures passing the IQ test, then biases conveyed by slurs like 'kraut' or 'hurray', as devices of conventional implicature, should pass the IQ test as well. Yet reports like:

(13) John said that Fritz is a kraut.

(14) John said that hurray Fritz got promoted.

do not seem to pass the IQ test. With (13) the narrator does not automatically report John's negative attitude toward Germans. Yet, it guarantees the offense by the reporter, i.e. the latter endorses/expresses the negative attitude toward Germans (Anderson & Lepore 2013). The slur scopes out of the report. A non-racist narrator would prevent him/herself from using the slur in his/her report. She would rather say something like: "I do not endorse any prejudice against Germans, but John said that Fritz is a kraut". Yet, Anderson & Lepore (2013) claim that even in a report like this the negative attitude toward Germans slips out of the report. As such, a report may be transparent concerning the way the attributee referred to the relevant class picked out by the slur appearing within the context of a report; it could thus fail to attribute the use of the slur to John. I leave to the reader to decide whether the N-word in John Lennon's famous "Women are the niggers of the world", or in some other metaphorical uses, comes out to be interpreted as a derogatory word. Be that as it may, the important lesson to bring home is that, since slurs do not pass the IQ test, they cannot be treated as conventional implicatures. As for the bias introduced by 'hurray' in (7) it cannot be reported as attributing to the attributee a favorable

---

within the scope of the that-clause. If it were to take wide scope, Bach's test would not work. People's intuitions may differ on whether 'but' takes narrow or wide scope. Yet, insofar as the narrow scope reading is not ruled out, the IQ test is still reliable.

[14] The semantic contribution of ACIDs can thus be compared to the semantic contribution made by parentheticals or non-restrictive relative clauses like e.g.: "John (Jane's husband) has been promoted" and "John, Jane's husband, has been promoted".

disposition or attitude toward John's promotion. As it stands, (14) is ungrammatical and, therefore, (7) does not pass the IQ test. Notice that if in a report 'hurray' scopes out, as in

(15)   Hurray, John said that Fritz got promoted.

it would express the positive bias of the reporter, not the one expressed by John in his uttering of (7). It would express, in other words, the narrator's positive attitude toward John saying that Fritz got promoted, not toward Fritz being promoted. Expressions like 'hurray' and 'alas'—that, following Predelli, have a null character—can be viewed as utterance modifiers like 'moreover' and 'in other words'. Like the latter, they do not pass the IQ test.[15] Hence, they do not contribute to what is said (they are merely communicational devices):

(16)   a.   Jane: "*Moreover*, Jeff is honest".
       b.   *Jane said that *moreover* Jeff is honest.
(17)   a.   Jane: "*In other words*, Fritz is honest".
       b.   *Jane said that *in other words* Fritz is honest.

Additions like 'moreover' and 'in other words' are vehicles to perform second-order speech acts (Bach 1999). And since they do not pass the IQ test, they do not contribute to what is said, they do not affect the proposition expressed, i.e. what I will characterize in the next section, following Perry (2002/12) and Korta & Perry (2011), as the official content of the utterance.

In short, it seems that slurs (and biases in general) cannot be dealt with by characterizing them as devices triggering conventional implicatures.

4 Biases *qua* Generalized Conversational Implicatures

One way to characterize slurs would be to adopt yet another important distinction proposed by Grice, *viz.* the distinction between particularized and generalized conversational implicatures, and treat them along the lines of the latter (Vallée 2014). In this way, we can argue that biases, being *conventionally* conveyed by the word used can be accommodated as instances of generalized conversational implicatures:

> Sometimes one can say that the use of a certain form of words in an utterance would normally (in the absence of special circum-

---

[15] The IQ test may be a good strategy to characterize what Predelli considers expressions with null character.

stances) carry such-and-such an implicature or type of implicature. Noncontroversial examples are perhaps hard to find, since it is all too easy to treat a generalized conversational implicature as if it were a conventional implicature. ... Anyone who uses a sentence of the form *X is meeting a woman this evening* would normally implicate that the person to be met was someone other than X's wife, mother, sister, or perhaps even close platonic friend. (Grice 1967/87: 37)

Following Grice, a generalized conversational implicature can be presumed or presupposed independently of a particular context of the utterance.[16] It constitutes, so to speak, the *default interpretation* a speaker would associate to an utterance. As such, it constitutes information that a competent speaker, *ceteris paribus*, computes given the linguistic input. Yet, being a *conversational* implicature, a generalized conversational implicature is cancelable:

Since, to assume the presence of a conversational implicature, we have to assume that at least the Cooperative Principle is being observed, and since it is possible to opt out of the observation of the principle, it follows that a generalized conversational implicature can be canceled in a particular case. It may be explicitly canceled, by the addition of a clause that states or implies that the speaker has opted out, or it may be contextually canceled, if the form of utterance that usually carries it is used in a context that makes it clear that the speaker is opting out. (Grice 1967/87: 39)

If, at his garden party, John asks Jane "Should we bring out more beer?" and Jane replies: "Most of the guests already left" the particularized conversational implicature is that they should not bring out more beer, while the generalized conversational implicature would be that not all the guests left.[17] Both implicatures can be cancelled. To cancel the particularized conversational implicature Jane could add "But all the beer-drinkers are still here". To cancel

---

[16]For a book-length, detailed discussion of generalized conversational implicatures, see Levinson (2000). See also Huang (2000). Among some (relatively) clear examples of generalized conversational implicatures we can mention that "Jane thinks there is a meeting tonight" implicates "Jane *does not know for sure* that there is a meeting tonight"; "Jane has two children" implicates "Jane has *no more than* two children"; "Jeff broke a leg" implicates "Jeff broke *his own* leg".

[17]If in the very same situation Jeff asked "What time is it?" and Jane replied "Most of the guests already left" the generalized conversational implicature would be the same, i.e. that not all the guests left, while the particularized one would be that it is late.

the generalized conversational implicature that not all the guests left she could add "Actually all the guests left". To be sure, to cancel the generalized conversational implicature Jane must somewhat flout the maxim of quantity, for she would say more than is required. If biases are classified as devices triggering generalized conversational implicatures they should be cancellable as well. Yet, with the utterance of a sentence like "I have no prejudices against Germans but Fritz is a kraut" one does not seem to be cancelling his/her negative bias toward Germans. If someone utters "I love wop " or "Wops are my favorite people" one fails to erase the negative attitude she manifests *vis-à-vis* Italians. The least we can say is that the negative attitude conveyed by a slur cannot be cancelled without difficulties. We need strong contextual inputs or quite articulated clauses to convey the fact that the speaker does not endorse the generalized conversational implicature usually triggered by the slur. It is open to question whether in a humorous sketch an actor dressed like an Italian pimp distributing drugs to the clients of his protégées, using 'wop' conveys a negative attitude toward Italians (Anderson & Lepore 2013 would claim that even in such a situation the slur slips out of the play and conveys a negative attitude toward Italians). A possible way to consider slurs (and other biases) along the lines of generalized conversational implicatures is to consider the latter as belonging to a theory of utterance(-type) meaning:

> The theory of GCI [generalized conversational implicatures] is not of course a theory of conversational idioms, clichés, and formulae, but it is *a generative theory of idiomacity*—that is, a set of principles guiding the choice of the right expression to suggest a specific interpretation and, as a corollary, a theory accounting for preferred interpretations... The theory thus belongs to the intermediate level of a theory of communication, the layer of utterance(-type) meaning. (Levinson 2001: 24; italics in the original)

The basic idea underlying this theory of communication focusing on generalized conversational implicatures must admit "the general contribution of a level at which sentences are systematically paired with preferred interpretations" (Levinson 2001: 27) Translated into Predelli's terminology this may amount to stating that we need a theory of biases where a given expression comes equipped with its character and the bias selecting the context of use. The theory of use would thus come close to the theory of generalized conversational implicatures *qua* a theory that systematically pairs expressions with preferred interpretations. To illustrate this, let us consider an utterance of:

(18)   Alas, Jane is a whore.

A competent speaker of English will compute, without further information (linguistic and/or contextual), that the speaker has *both* (i) a negative attitude toward female sex-workers, triggered by the slur-word 'whore' instead of the neutral 'sex-worker' *and* (ii) a negative attitude regarding Jane being a sex-worker, triggered by the interjection 'alas'. That is, an utterance of (18) encompasses two biases, one triggered by the slur 'whore' regarding sex-workers and the other triggered by the interjection 'alas', regarding Jane being one of them. These are cases of alleged non-truth-conditional meaning. If instead of (18) the speaker or writer of had produced:

(19)   Jane is a sex-worker.

she would have said (in Grice's sense) the same thing, i.e., she would have expressed the same proposition (or Kaplanian content), i.e. that Jane is a sex-worker. Yet, in expressing this content the speaker/writer conveys different attitudes. While (19) is neutral, (18) conveys the speaker/writer's biases; the latter triggers two generalized conversational implicatures.

5 Biases and Reflexive Contents

I now turn to the pluri-propositionalist theory of communication that, if I am right, encapsulates the gifts of Predelli's theory of use and the ones of a theory of generalized conversational implicatures. The theory of communication I have in mind attempts to take into account the different motivations guiding a speaker's selection of a given utterance over another expressing the same content, i.e. what motivates a speaker to say something in a particular way instead of another way. The proposal I have in mind is Perry's critical referentialism.[18] Perry's underlying idea is that:

> I cannot accept that a semantic theory can be correct that does not provide us with an appropriate interface between what sentences mean, and how we use them to communicate beliefs in order to motivate and explain action. A theory of linguistic meaning should provide us with an understanding of the properties sentences have that lead us to produce them under different circumstances, and react as we do to their utterance by others. (Perry 2012: 9)

---

[18] See Perry (2001/2012) and Korta & Perry (2011).

On this account the utterance of a sentence comes equipped with various contents. The basic idea is that every utterance is *systematically* associated with a family of contents that derive from the combination of the speaker's plan (his/her intentions and beliefs), the conventions exploited, and the circumstances of the utterance. The truth-conditions of these contents can be classified in different ways; some of these contents are reflexive or utterance-bound contents, with the utterance itself as a constituent. In other words:

> Any statement, whether or not it contains indexicals, has multiple reflexive contents associated with it, which will be grasped by a semantically competent listener and are necessary for an account of cognitive significance. ... if we examine carefully what the problems that cases of co-reference and no-reference pose for semantic theory, we shall see that these problems can be solved at the level of reflexive content. (Perry 2012: 12-13)

These variegated contents expand from the purely reflexive to the referential, or official, content: they constitute a family of gradually less reflexive and less contextually dependent contents. The content expressed by the utterance is, according to Perry, usually considered the referential content or official content and corresponds to the intuitive direct-reference concept of what is said (or Kaplanian content).

To illustrate the difference between official and reflexive contents, let's suppose that Jane utters (20) while David utters (21):

(20)   I love champagne

(21)   Jane loves champagne

From the direct reference standpoint, the proposition expressed by (20) and (21) is the following:

(20/21)a.   That **Jane** loves champagne[19]

(20) and (21) express the same singular proposition, with Jane and the property of loving champagne as constituents. (20) and (21) have the same official content.

This, though, does not explain why, in normal circumstances, if Jane's aim is to get John to invite her to drink some champagne, she would utter (20) but

---

[19] I follow Perry's notation of using boldface for content constituents; roman when the constituent is the object designated; *italics* **when it is not an object but an identifying condition**.

not (21). John may even be unaware of Jane's name. Even if he knew her name it would be awkward for Jane to speak as Maradona and De Gaulle used to do, and utter (21).

Perry's account starts from distinguishing the reflexive content of an utterance like (20).[20] The indexical (reflexive) content of (20) corresponds to what a hearer would understand, given his/her knowledge of English and no other contextual information besides the fact that (20) has been produced. This can be rendered by the following:

(20)   b.   That ***the speaker of*** **(20)** loves champagne

(20b) is a reflexive content, for it has (20), the utterance itself, as a constituent: (20b) is about (20) itself. (20b) is a proposition associated with (20) in virtue of the meaning of the sentence, the type, "I love champagne". This property is given by the linguistic meaning of the type, independently of any contextual feature. Any competent speaker of English grasps this meaning, i.e. grasps (20b). If one were to read it on a blackboard or in an anonymous letter without knowing who wrote it (i.e. who the referent of 'I' is) a competent speaker would understand at least (20b). (20b), though, is not the content expressed, *viz.* it is not what Jane said in uttering (20). Jane said that she loves champagne; she didn't say that the speaker of (20) loves champagne. To borrow Frege's terminology (20b) is not the subject matter of (20). Jane talked about herself (the subject matter); she didn't talk about the words she used. Yet (20b) plays a crucial role in Jane's communicative plan. Given that John understands (20) and notices that the speaker of (20) is the person sitting next to him, he can grasp the following:

(20)   c.   That ***the person next to me*** loves champagne

This is not the content expressed by (20); it is not what Jane said, but it is the relevant content that John should grasp for Jane's plan to succeed—that is, to

---

[20] The distinction between reflexive truth-conditions or reflexive content and incremental truth-conditions or official content bears a similarity to and is inspired by Reichenbach's (1947) token-reflexive treatment of indexical expressions. According to Reichenbach, the meaning of an indexical such as 'I' is explained using a reflexive description of the form 'the utterer of this token' while the meaning of 'you' is explained as 'the addressee of this token'. This conception was refined with Kaplan's (1977) content/character distinction. The character (or linguistic meaning) of an indexical can be represented as a function taking as its argument the context and delivering as its value the content or referent. Thus, the character of 'I' can be represented as 'the agent of this utterance' and the character of 'you' as 'the addressee of this utterance'. In short, the character or linguistic meaning of an indexical is what a competent speaker/hearer masters.

trigger John to invite her to drink champagne. It is this content and not the indexical or the referential content that accounts for the cognitive motivation and cognitive impact of (20). (20c) is understood on the basis of the hearer's knowledge of English, i.e. (20b), and perceptual awareness of the context of utterance, who the speaker is and where she is located. John grasps it through his perception of the utterance.

With proper names the situation is slightly different. The reflexive content of (21), uttered by David to John, would be the following (cf. Perry 2012: 122):

(21) b. That *the person the convention exploited by* **(21)** *permits one to designate with 'Jane'* loves champagne.

John's mastery of English does not give him more than that. Without any hints about the naming convention the speaker is using to refer to a particular person, John is not in the position to grasp the official content of (21), i.e. he is cognitively blind about the content expressed. Furthermore, John will not get any other "intermediate" content that could give him an idea about the referent picked out by David's use of 'Jane'. While (20b) directs the hearer, John, to a particular aspect of the context of the utterance, the speaker, (21b) does not lead John to any contextual aspect and, thus, to further less context-sensitive (or incremental, to use Perry's terminology) contents. In a nutshell, while (20c) allows John to invite Jane to drink champagne, (21b) does not prompt John to invite Jane to drink champagne. Even if John happens to know the person called 'Jane' that David is referring to, so that he understands what David is saying—i.e. he grasps the official content of (20/21a)—this is not enough to trigger John to invite Jane to drink champagne: to do so, he must recognize Jane as the person sitting close to him, *viz.* he needs to grasp (20c).

An utterance's reflexive content helps us to deal with problems pertaining to cognitive significance. As Frege (1892) pointed out, one expands one's knowledge when one comes to realize that Hesperus is Phosphorus. For, although the referential or official content of "Hesperus is Phosphorus" does not differ from the one of "Hesperus is Hesperus", their reflexive contents differ.[21]

---

[21] As Kaplan suggested in his seminal "Demonstratives" (1977), to deal with the notion of cognitive significance we have to focus on the expressions' character (roughly, linguistic meaning). Thus the cognitive significance of an utterance of 'My pants are on fire' differs from the cognitive significance of 'His pants are on fire' insofar as the character of 'my' differs from the character of 'his' even in the case the two expressions are co-referential. One utterance may trigger different behaviors from the other. Inspired by Kaplan, Perry argues that problems pertaining to cognitive significance ought to be explained at the level of reflexive truth-conditions: "(a) The cognitive significance of an utterance S in language L is a semantic property of the

I will now try to show how Predelli's theory of bias fits within the Perry-inspired account of communication.²² The reflexive truth-conditions of (18) and (19) that I repeat here:

(18)   Alas, Jane is a whore

(19)   Jane is a sex-worker

could be cashed out, respectively, as follows:

(18/19)a.   That *the person the convention exploited by* **(18)** *permits one to designate with 'Jane'* **is a sex-worker**.

while its official content would correspond to the proposition

(18/19)b.   That **Jane** is a sex-worker.

Yet, since (18), unlike (19), also conveys the biases that the utterer (or writer) has a negative attitude both regarding sex-workers and regarding the fact that Jane is a sex-worker, the reflexive content of (18) also encapsulates the agent of (18)'s biases, i.e.

(18)   c.   That *the agent of* **(18)** **has a negative attitude toward** *the individual s/he refers to using 'Jane'* **being a sex-worker**

(18)   d.   That *the agent of* **(18)** **has a negative attitude toward sex-workers**

(18c) should capture the generalized conversational implicature that the agent of (18) has a negative attitude (conveyed by the use of the interjection 'alas') concerning Jane being a sex-worker, while (18d) the generalized conversational implicature that the agent of (18) has a negative attitude, triggered by the slur 'whore', toward sex-workers in general. If Perry is right in arguing that reflexive contents help us to classify agents' attitudes (and, as such, to deal with problems pertaining to cognitive significance), then the reflexive contents

---

utterance. (b) It is a property that a person who understands the meaning of S in L recognizes. (c) The cognitive significance of an utterance of S in L is a proposition. (d) A person who understands the meaning of S in L, and accepts as true an utterance of S in L, will believe the proposition that is the cognitive significance of the utterance. (e) A person who understands the meaning of S in L, and sincerely utters S, will believe the proposition that is the cognitive significance of his utterance" (Perry 1988: 194).

²²In a recent paper Vallée (2014) convincingly proposes a similar (Perry-inspired) account concerning slurs. One of the aims of my paper is to show how Vallée's account generalizes to other categories of biases.

triggered by the interjection 'alas' and the slur 'whore' capture the attitude of the agent of (18).

The picture I am proposing, it seems to me, comes close to the view championed by Predelli's account. In particular, it comes close to Predelli's view that the meaning of an expression can be considered as a pair comprising the character of the given expression *and* the bias it conveys. Thus 'whore' and 'sex-worker' will contribute the same property into the proposition expressed (the official content), while differing in their contribution into the reflexive content. The interjection 'alas' does not make a contribution into the official content; its contribution affects the agent's attitude *vis-à-vis* the latter and it is captured at the reflexive level by (18d).

Furthermore, if we take on board Grice's view on generalized conversational implicatures as being the information that one extracts from, say, an anonymous letter, we can deal with the conventional aspect associated with slurs and interjections—that is, the information that a competent speaker can infer in virtue of his/her mastery of the language. This also allows us to capture Predelli's view that if we fix the context of use some sentences are self-verifying. In the context of a face-to-face communication, for instance, an utterance of

(22)   I exist [said by John]

is self-verifying. Its self-verifiability is not triggered by the official content, i.e. the proposition that John exists, but by its reflexive content, i.e.:

(22)   a.   That ***the agent of (22) exists when s/he produces it (at the time and location of the utterance)***.

This information ought not to be confused with particularized conversational implicatures. The speaker of (22) may *conversationally* implicate that he is a male, that she or he has a Californian accent, that she is an adult person, that she or he speaks fluent English, and so on and so forth. After all, the audience must perceive the utterance. The relevant information concerning the self-verifiability of an utterance like (22) is *conventionally* imparted insofar as one *cannot* utter (22) and contradict oneself. If (22) were used in a will it would force a switch of context (it would be, as Predelli states, an obstinate use of the indexical). As competent speakers, the heirs, on hearing "I do not exist" would easily compute that the agent introduced by the signature does not exist at the time they read/hear the utterance. This competence is captured at the reflexive level of content.

6 Conclusion

After this excursion toward Predelli's theory of bias and his account of communication, I hope to have shown that many of the data he proposes can be dealt with within a pluri-propositionalist theory of communication that, to my understanding, is independently motivated—*viz.* it is not an *ad hoc* theory motivated by the presence of various biases in natural languages. Yet it can be developed so as to account for the rich amount of data that Predelli's theory of bias elegantly deals with. In particular, I attempted to show how Predelli's view that an expression comes equipped with both a character and a bias can be accommodated in distinguishing between the official content and the reflexive one. While the character may contribute to the expression of the official content (what is said), the bias, along the lines of generalized conversational implicatures, contributes to the reflexive content. As I see it, Predelli's theory of biases, and his general account of communication, fits nicely within Perry's and Korta & Perry's account of communication, though I am not sure the latter will agree with my understanding and development of their original framework, and, in particular, whether they sit with me in arguing that generalized conversational implicatures contribute into the reflexive content of an utterance.

References

Anderson L. & E. Lepore. (2013). "Slurring Words". *Noûs* 47(1): 25–48.

Bach K. (1999)."The Myth of Conventional Implicatures". *Linguistic & Philosophy* 22: 327–66.

Cappelen H. & Lepore E. (2005). *Insensitive Semantics*. Oxford: Blackwell.

Carston R. (2004). "Relevance Theory and the Saying/Implicating Distinction". In L. Horn & G. Ward (eds) *Handbook of Pragmatics*. Oxford: Blackwell: 607–632.

Corazza E. (2002). "Temporal Indexicals and Temporal Terms". *Synthese* 130(3): 441–460.

— (2004). *Reflecting the Mind: Indexicality and Quasi-Indexicality*. Oxford: Oxford University Press.

— (2012). "Same-Saying, Pluri-Propositionalism, and Implicatures". *Mind & Language* 27(5): 546–69.

Corazza E., Fish W. and Gorvett J. (2001). "Who is I?". *Philosophical Studies* 107(1): 1–21.

Frege G. (1892). "On Sense and Meaning". In P. Geach & M. Black (eds). (1952). *Translations from the Philosophical Writings of Gottlob Frege.* Oxford: Blackwell: 56–78.

Grice P. (1967/87). "Logic and Conversation". In P. Grice (1989). *Studies in the Way of Words.* Cambridge Mass: Harvard University Press

Huang Y. (2000.) *Anaphora: A Cross Linguistic Study.* Oxford: Oxford University Press.

Kaplan D. (1977). "Demonstratives". In J. Almog, J. Perry, and H. Wettstein (eds) (1989). *Themes from Kaplan.* Oxford: Oxford University Press: 481–463.

Korta K. & Perry J. (2011). *Critical Pragmatics.* New York: Cambridge University Press.

Levinson S. (2000). *Presumptive Meaning: The Theory of Generalized Conversational Implicatures.* Cambridge MA: MIT Press.

Perry J. (1986). "Thoughts without Representation". *Proceeding of the Aristotelian Society* 60: 137–52. Reprinted in J. Perry (2000). *The Problem of the Essential Indexical and Other Essays.* Palo Alto CA: CSLI Publications: 171–88.

— (1988). "Cognitive Significance and the New Theory of Reference" *Noûs* 22: 1-18. Reprinted in J. Perry 2000. *The Problem of the Essential Indexical and Other Essays.* Palo Alto CA: CSLI Publications: 189–206.

— (2002/12). *Reference and Reflexivity* (second ed.). Palo Alto CA: CSLI Publications

Potts C. (2007). "The Expressive Dimension". *Theoretical Linguistics* 33: 165–198.

Predelli S. (1998). "I am not Here Now". *Analysis* 58(2): 107–15.

— (2013). *Meaning without Truth.* Oxford: Oxford University Press.

Reichenbach H. (1947). *Elements of Symbolic Logics.* New York: Free Press.

Sidelle A. (1991). "The Answering Machine Paradox". *Canadian Journal of Philosophy*, 81(4): 525–539.

Smith Q. (1989). "The Multiple Uses of Indexicals". *Synthese* 78: 167–191.

Vallée R. (2014). "Slurring and Common Knowledge of Ordinary Language". *Journal of Pragmatics* 61: 70–90.

Vision G. (1985). "'I am Here Now'". *Analysis* 45(4): 198–199.

Williamson T. (2009). "Reference, Inference and the Semantics of Pejoratives". In J. Almog & P. Leonardi (eds). *The Philosophy of David Kaplan.* Oxford: Oxford University Press: 137–158.

Chapter 8
Metaphor, Relevance and the Interpretation of Fiction*

Deirdre Wilson

Abstract In a series of works that raise intriguing questions about the interpretation of fiction, Anne Reboul (1986, 1987) asks "why people are prepared to spend time reading or listening to fiction, even though they know it is false". From the perspective of relevance theory, where the allocation of attention is seen as guided by expectations of relevance, this is a special case of a more general question: How can an utterance be relevant without being true? The answer Reboul proposes is that fictional utterances might achieve relevance in similar ways to metaphor, another type of utterance widely seen as capable of being relevant without being true. However, the account of metaphor that Reboul has in mind—first put forward by Dan Sperber and Deirdre Wilson (1986a/1995)—has evolved considerably over the years, and in more recent versions, metaphors are no longer seen as capable of being relevant without being true. In this paper, I defend this more recent "ad hoc concept" account of metaphor against reservations expressed by Reboul (2011), and explore alternative ways of preserving her original insights into what makes fiction worth reading or listening to even though we know it is false.

Keywords Explicature · Implicature · Ad hoc concept · Weak communication · Figurative language · Non-propositional effects

---

*I first met Anne Reboul at the LSA Summer School in Los Angeles in 1983, and have valued her warmth, wit and friendship ever since. This paper is a continuation of our many enjoyable conversations.

1 Introduction

In a number of works that raise important questions about the nature and interpretation of fiction,[1] Anne Reboul asks "why people are prepared to spend time reading or listening to fiction, even though they know it is false" (Reboul 1987: 729). From the perspective of relevance theory, where the allocation of attention is seen as guided by expectations of relevance, this is a special case of a more general question: How can an utterance be relevant without being true?

Reboul (1986, 1987) sees fictional utterances as "very near to metaphor", another type of utterance widely treated as capable of being relevant without being true:

> Like metaphor, [fictional discourse] need not purport to represent an actual state of affairs as long as it achieves optimal relevance: that is, as long as it gives rise to sufficient effects which describe actual states of affairs (the "message" of the work) for the minimum necessary processing effort. (Reboul 1987: 729)

In defending this view, she draws on an early relevance theory account of metaphor which treats a metaphorical utterance such as (1) as having no explicature, and hence as making no true assertion, or expressing no proposition that the speaker intends to endorse:

(1) The house is in a *magical* spot[2]

Instead, the metaphorically used term is seen as activating an array of implications—e.g. that the house is in a spot that is beautiful, delightful, remarkable, extraordinary, outstanding, astonishing (etc.) in a way rarely found in ordinary life—some subset of which the speaker does intend to endorse, thus helping to satisfy the hearer's expectations of relevance (Sperber and Wilson 1986a, 1986b). On this approach, metaphors are prime examples of utterances which can be relevant without being true.

As a result of developments in the relatively new field of lexical pragmatics, this early relevance theory account of metaphor has evolved in several ways.[3] In particular, a metaphorical utterance such as (1) is no longer seen as

---

[1] See for instance Reboul (1986, 1987, 1990, 1992, 1997, 2000, 2001, 2008, 2009, 2011, 2012).

[2] As a result of repeated metaphorical use, the term *magical* is likely to acquire an additional lexicalised meaning, at least for some speakers. Pragmatics sheds light on the processes by which these additional meanings are acquired.

[3] See for instance Wilson & Carston (2006, 2007); Sperber & Wilson (2008).

lacking explicatures, and hence as incapable of being true. Instead, the concept encoded by a metaphorically used term is seen as undergoing pragmatic "adjustment", or "modulation" in the course of the comprehension process, with the resulting "ad hoc" (unlexicalised) concept contributing to the explicature of the utterance, and hence to its truth conditions.[4] Thus, *magical* in (1) might be interpreted as conveying an ad hoc concept, MAGICAL*, which shares some of its implications with the encoded concept MAGICAL and figures in the explicature of (1). On this approach, metaphors are no longer prime examples of utterances that can be relevant without being true, and the suggested parallel between fictional discourse and metaphor breaks down.

Of course, the parallel breaks down only on the assumption that the "ad hoc concept" account of metaphors is more adequate than the earlier "no explicature" account. Reboul (2011) argues that the "ad hoc concept" account loses many of the advantages of the earlier account. Her reservations, if justified, would have implications not only for the analysis of metaphor itself but also for the suggested parallel between metaphor and fiction. In this paper, I will defend the "ad hoc concept" account against Reboul's reservations, and consider some alternative ways of preserving her original insight that fictional utterances can be relevant without being true.

2 Metaphor and relevance theory

Relevance theory's approach to metaphor – in both the earlier and the later versions – rests on a number of basic claims. The first is that utterance comprehension is not guided by a presumption of literal truthfulness but by a presumption of optimal relevance (Wilson & Sperber, 1981, 2002). This is a consequence of the more basic claim that utterances do not *encode* the speaker's intended meaning, but merely provide *evidence* of that meaning. It follows that, in interpreting the word *magical* in (1), there is no presumption that the speaker is intending to communicate the encoded concept MAGICAL itself; instead, she may be speaking loosely and merely intending to communicate some of the implications activated by the use of this concept in that particular context: for instance, that the spot is beautiful, delightful, remarkable, extraordinary, outstanding, astonishing (etc.) in a way rarely found in ordinary life. A subset of these implications would then be among the implicatures of (1), and might help to make it relevant in the expected way. Instead of a presumption of literal

---

[4]For approaches to lexical adjustment from different perspectives, see Recanati (1995, 2004), Carston (1997, 2002, 2010, 2015), Sperber & Wilson (1998, 2008), Bezuidenhout (2002), Wilson & Sperber (2002), Wilson & Carston (2006, 2007), and Vega Moreno (2007).

truthfulness, there is a presumption of optimal relevance; instead of the claim that every utterance encodes a thought of the speaker's, there is the claim that every utterance is a more or less literal interpretation of the thought the speaker intends to convey.

An important aspect of relevance theory's approach to metaphor is the claim that there is a continuum of cases of loose use, ranging from the merest approximation at one end to the most poetic metaphor at the other, with no principled dividing line between them. As Sperber and Wilson (2008: 84–5) put it,

> We see metaphors as simply a range of cases at one end of a continuum that includes literal, loose and hyperbolic interpretations. In our view, metaphorical interpretations are arrived at in exactly the same way as these other interpretations. There is no mechanism specific to metaphors, no interesting generalisation that applies only to them. In other terms, linguistic metaphors are not a natural kind, and "metaphor" is not a theoretically important notion in the study of verbal communication.

According to this "deflationary" approach, metaphor and hyperbole are understood in exactly the same way as more mundane cases of loose use which would not normally be classified as figurative. In (2), for instance, *kill*, *glass* and *water* would all be seen as cases of loose use, although *kill* is traditionally classified as a hyperbole and *glass* and *water* are not:

(2)   I could *kill* for a *glass* of *water*.

On this approach, all three loose uses would be interpreted as conveying not the concepts they encode—KILL, GLASS and WATER—, but only some of the implications activated by those concepts in that particular context. These basic features of relevance theory's approach to metaphor have remained constant over the years.

However, the claim made in the early relevance theory account that loose, hyperbolic or metaphorical uses make no contribution to explicatures proved increasingly hard to maintain. In the first place, it entailed that vast swathes of ordinary non-figurative utterances such as (3) and (4), which most people would accept as true (or "true enough"), are not only incapable of being true but do not explicitly communicate anything at all:

(3)   Oslo is *north* of London

(4)   The world is *round*

Neither (3) nor (4) is strictly and literally true. In frameworks with a norm or maxim of literal truthfulness, it is customary to distinguish between "minor" infringements such as these, which are handled by appeal to "contextually determined standards of precision" in telling the truth, and the "major" violations found in figurative utterances, which are handled in an entirely different way. Given its "deflationary" claim that there is a continuum of cases between loose use, hyperbole and metaphor, with no principled dividing line between them, this solution is not available to relevance theory.[5] Either loose use, hyperbole and metaphor must all be seen as capable of being true (or "true enough") or none can.

In fact, the standard embedding tests suggest that loose use, hyperbole and metaphor do affect the truth conditions of utterances in which they occur. Consider (5) and (6):

(5)   If Oslo is *north* of London, the weather there must be very cold.

(6)   If you find a house in a *magical* spot, you should do your best to buy it.

The condition in which the speaker of (5) is claiming that the weather in Oslo must be very cold is not that Oslo is due north of London (as a presumption of literal truthfulness would predict), but that Oslo is in an approximately northerly direction from London. Similarly, the condition in which the speaker of (6) is encouraging the addressee to buy a house is not that he finds one in a spot that is literally celestial or blessed, but that he finds one in a spot that is beautiful, delightful, remarkable, extraordinary, outstanding, astonishing (etc.) in a way rarely found in ordinary life. In other words, the fact that *north* and *magical* are used loosely rather than literally contributes to the explicatures of (5) and (6), and therefore affects their truth conditions. How these cases should be analysed on the "no explicature" account was never entirely clear.[6]

A more serious problem pointed out by Robyn Carston (1997) is that the denotation of a lexical item can undergo simultaneous broadening and narrowing in the course of the comprehension process. From the start, relevance theorists saw lexical narrowing as one of the many varieties of pragmatic enrichment that contribute to explicatures, and therefore affect the truth conditions of utterances. If lexical items undergo simultaneous broadening and narrowing, it is hard to maintain that narrowing contributes to explicatures while broadening

---

[5]For a detailed critique of standard treatments and a defence of the deflationary approach, see Wilson & Sperber (2002), Sperber & Wilson (2008)

[6]I recall Stephen Levinson using examples similar to (6) in a draft paper on generalised implicatures circulated in the late nineteen-eighties, and wondering how they should be dealt with on a "no explicature" account.

does not. Consider (7):

(7)   a.   Shall we invite Camilla to the children's party?
           b.   She's a *rottweiler*.

In (7b), *rottweiler* is metaphorically used to implicate that Camilla is aggressive, unfriendly, a potential risk to children (etc.), and therefore not a good person to invite to a children's party. In order to derive these implicatures, the hearer must assume that the speaker is (a) broadening the denotation of *rottweiler* to include both people and rottweilers, and (b) narrowing it to include only that subset of people and rottweilers which are aggressive, unfriendly, a potential risk to children (etc.). On the further assumption that Camilla belongs to this newly created category—call it ROTTWEILER*—, he is then justified in inferring that Camilla is not a good person to invite to a children's party. For the inference to be sound and the implicature properly warranted, the explicature of (7b) must be seen as containing the concept ROTTWEILER*, constructed ad hoc in order to satisfy expectations of relevance.[7] On this approach, the speaker of (7b) would be analysed as communicating not the encoded concept ROTTWEILER but the "ad hoc" (unlexicalised) concept ROTTWEILER*. On this "ad hoc concept" account, the problems raised by the earlier "no explicature" account disappear.[8]

3   Anne Reboul on the "ad hoc concept" approach

A long-standing goal of relevance theory has been to account for what Sperber & Wilson (1986a/1995) call "vague communication", a type of communication often found with figurative utterances and on the implicit side of the explicit-implicit divide. According to Sperber & Wilson (1986a/1995: 56–7),

> We all know, as speakers and hearers, that what is implicitly conveyed by an utterance is generally much vaguer than what is explicitly expressed, and that when the implicit import of an utterance is explicitly spelled out, it tends to be distorted by the elimination of this often intentional vagueness. The distortion is even greater in the case of metaphor and other figures of speech, whose

---

[7]The term *ad hoc concept* was suggested by the work of Lawrence Barsalou on ad hoc categories (e.g. Barsalou 1987), which reveals the remarkable flexibility of the human mind in constructing appropriate categories to suit different contexts.

[8]On the "ad hoc concept" approach to lexical pragmatics, see Wilson & Carston (2007), Carston (2015), Wilson & Kolaiti (2017).

poetic effects are generally destroyed by being explicitly spelled out... We see it as a major challenge for any theory of communication to give a precise description and explanation of its vaguer effects.

In attempting to meet this challenge, Sperber & Wilson reject the standard view that what is communicated by an utterance as invariably a *speaker's meaning* consisting of a single proposition, or at most a small set of propositions. Instead, they treat the intended import of an utterance as a *set*, or *array*, of propositions which may vary indefinitely in size. At one extreme, the intended import may consist of a single proposition or a small set of propositions, as on standard accounts; at the other, it may consist of a vast array of propositions; and there is a continuum of cases in between (Sperber & Wilson 2015).

Among the array of propositions communicated by an utterance, some may be more strongly communicated than others. A proposition is *strongly* communicated to the extent that the speaker's intention to convey it is both salient and strongly evidenced – or, in relevance theory terms, to the extent that the speaker's intention to convey it is strongly *manifest*.[9] The more strongly manifest the speaker's intention to convey a specific proposition, the more strongly it is communicated. For a hearer of (7b) who is familiar with the cultural stereotype of a rottweiler, the implicature that Camilla should not be invited to the children's party would be rather strongly communicated, since it answers question (7a) and is easily derivable from the assumption that Camilla has any of the stereotypical properties made salient by the encoded concept ROTTWEILER. By contrast, the assumption that Camilla has any particular one of those properties (i.e. that she is aggressive, or unfriendly, or a potential risk to children, etc.) would be rather less strongly communicated, since the speaker's intention to communicate that specific proposition is less strongly manifest. This introduces an element of vagueness into the interpretation of (7b), so that a hearer who concludes, for instance, that Camilla is a potential risk to children must take some of the responsibility for accepting it himself.

With a poetic metaphor such as *Juliet is the sun,* the interpretation is much vaguer. Here is a widely quoted interpretation of this metaphor put forward by the philosopher Stanley Cavell (1965/1976: 78–9):

> Romeo means that Juliet is the warmth of his world; that his day begins with her; that only in her nourishment can he grow. And his declaration suggests that the moon, which other lovers use as

---

[9]On manifestness, see Sperber & Wilson (1986a/1995: 39–46; 2015).

emblem of their love, is merely her reflected light, and dead in comparison, and so on. ... The 'and so on' which ends my example of paraphrase is significant. It registers what William Empson calls 'the pregnancy of metaphors', the burgeoning of meaning in them.

On a relevance-theoretic approach, instead of a few fairly strong implicatures, the intended import of a poetic metaphor such as *Juliet is the sun* is seen as consisting of a vast array of weak implicatures (Sperber & Wilson 1986a/1995: 217–24; 2008). This helps to explain not only the vagueness of poetic metaphors but also the open-endedness of their interpretations—what Cavell calls "the burgeoning of meaning in them"—and the fact that these interpretations are likely to vary somewhat from one addressee to another. It also helps to explain another widely noted feature of creative metaphors: that they cannot be paraphrased without loss or distortion of some of their import. An explicit paraphrase strengthens and focuses attention on a small subset of the vast array of weak implicatures, while weakening or diverting attention from others. As the philosopher A.C. Bradley (1909: 19) puts it,

> In true poetry it is impossible to express the meaning in any but its own words, or to change the words without changing the meaning.

Reboul's main reservation about the "ad hoc concept" account is that it no longer helps to explain these widely noted features of creative metaphors:

> In the current account, the *implicature*-based analysis has been replaced by an *explicature*-based analysis... Even though there is nothing to prevent a metaphor from having several explicatures, it does not seem likely that the explicature-based account allows as much liberty of interpretation to the hearer as the vintage Relevance-Theoretic account did. In other words, it is not clear why the explicature is not a paraphrase. Thus, it is not clear either that the current account can explain why paraphrasing a metaphor leads to an interpretive loss. (Reboul 2011: 11)

If I have understood it correctly, her worry is twofold: first, that an explicature such as "Juliet is the sun*" fails to capture the open-endedness in the interpretation of creative metaphors and the fact that they cannot be paraphrased without loss; and second, that an explicature such as "Juliet is the sun*" cannot act as a starting point for deriving the vast array of weak implicatures described in the earlier "no explicature" account, which is either lost or incorporated into the ad hoc concept itself. She concludes that "there is no way, on the current ["ad hoc concept"] account, that metaphors can produce an array of weak implicatures" (Reboul 2011: 13).

This seems to me an unduly pessimistic conclusion, and not one that defenders of the "ad hoc concept" accounts proposed by Carston (1997, 2002, 2015), Sperber & Wilson (1998, 2008) or Wilson & Sperber (2002) would draw. All these writers endorse the key claims of the earlier "no explicature" account, that metaphors typically communicate a wide array of weak implicatures, that their interpretations are open-ended, and that they are therefore not paraphrasable without loss. Here is what Wilson & Sperber (2002, 617-8) say about the interpretation of poetic metaphors, for which they propose a rather detailed "ad hoc concept" account:

> The more metaphorical the interpretation, the greater the responsibility the hearer has to take for the construction of implicatures (i.e. implicit premises and conclusions), and the weaker most of these implicatures will be. Typically, poetic metaphors have a wide range of potential implicatures, and the audience is encouraged to be creative in exploring this range (a fact well recognized in literary theory since the Romantics). Communication need not fail if the implicatures constructed by the hearer are not identical to those envisaged by the speaker. Some freedom of interpretation is allowed for, and indeed encouraged, by those who speak metaphorically.

Rather than amounting to an explicit paraphrase, Sperber & Wilson see the explicature "Juliet is the SUN*" as derived by backwards inference from the vast array of weakly implicated conclusions activated in that context by the encoded concept SUN, thus introducing into the explicature an element of vagueness commensurate with the size of that array.

An important feature of the "ad hoc concept" account is the idea that in the course of comprehension, tentative hypotheses about explicatures, implicit premises (derived from the encyclopaedic entry of the encoded concept) and implicated conclusions are "mutually adjusted" with each other and with the expectation of relevance, so that the explicature, combined with the implicit premises, *warrants* the implicated conclusions. Consider (7b) again:

(7) a. Shall we invite Camilla to the children's party?
b. She's a *rottweiler*.

As noted above, to warrant the implicated conclusion that Camilla should not be invited to the party, the denotation of the term *rottweiler* has to be broadened to include both rottweilers and people, and narrowed to include only that subset of rottweilers and people whose properties make their presence at a children's party inadvisable. Here is an account of how the ad hoc concept might be derived, modelled on those proposed in Wilson & Sperber (2002):

> Mary uses the word *rottweiler* to indicate the concept ROTTWEILER*, which is part of what she wants to convey. Peter reconstructs this concept by treating the word *rottweiler* and its associated mental encyclopaedic entry as a source of potential implicit premises such as "Rottweilers are unfriendly", "Rottweilers are aggressive", "Rottweilers are a potential risk to children". From these implicit premises and a still-incomplete interpretation of Mary's explicit meaning, he tentatively derives the implicit conclusions that Camilla is unfriendly, aggressive and a potential risk to children, and is therefore not a good person to invite to a children's party. He then arrives by backwards inference at the full interpretation of the explicit content that Camilla is a ROTTWEILER* – where ROTTWEILER* is the meaning indicated by *rottweiler*, and is such that Camilla's being a ROTTWEILER* is relevant-as-expected in the context – and its constituent concept ROTTWEILER*.

Analysed in this way, lexical interpretation, like utterance interpretation in general, crucially involves a process of mutually adjusting explicit content, implicit premises and implicated conclusions so as to warrant the implicatures that satisfy the audience's expectations of relevance.[10] The fact that ad hoc concepts are derived by backwards inference from an array of weak implicatures introduces an element of indeterminacy into the explicature itself: the wider the array of implicatures, the weaker the explicature that results. Thus, the "ad hoc concept" account helps to explain just the same features of creative metaphors as the earlier "no explicature" account, while solving a number of problems raised by that account.

---

[10] This process is seen as taking place in parallel rather than in sequence, with tentative hypotheses about explicit content, implicit premises and implicated conclusions being continuously readjusted until a stable interpretation is reached.

4 Metaphor, relevance theory and "non-propositional" effects

The interpretation of figurative utterances is often seen as involving not only a propositional content but also a variety of "non-propositional" effects. A few early works on relevance theory suggested that these "non-propositional" effects might call for a new type of "rhetorical" mechanism, distinct from the inferential pragmatic mechanisms proposed in the theory. According to Wilson & Sperber (1981: 163–4),

> In addition to the propositions it expresses or implicates, an utterance may suggest to the hearer certain non-propositional lines of interpretation—for example by evoking images or states of mind—which are precisely characteristic of figurative utterances. ... What seems to be needed is a new type of interpretive mechanism, in addition to the semantic and pragmatic ones already available, which can account for irony, metaphor and figurative interpretation in general.

This idea was quickly dropped in favour of a fully inferential approach, but it is sometimes felt, by both supporters and critics of relevance theory, that the attempt to unify pragmatics with rhetoric may have left out something crucial. The importance of "non-propositional" effects is particularly keenly felt by those concerned with the analysis of literary works (see for instance Pilkington 2000, 2001). Reboul (2011) raises the question of how these effects can be accommodated in an explicitly inferential approach to pragmatics such as relevance theory.

Reboul (2007, 2011) points to the existence of neurolinguistic evidence suggesting that sensorimotor mechanisms are activated in the course of language comprehension (e.g. Hauk et al. 2004, Nazir et al. 2008). Literary scholars working on "embodied" cognition and communication draw on similar evidence in their analyses of poetry and fiction (e.g. Bolens 2012, 2016; Cave 2016; Cave & Wilson 2018). This evidence might be taken to support the claim that two types of pragmatic mechanism are needed for a full account of the interpretation of metaphor and fiction: one properly inferential and used in the derivation of explicatures and implicatures, the other non-inferential and used to create "non-propositional" effects.

For someone who sees inference as involving an "abstract" logic of the type used in conscious syllogistic reasoning, the case for invoking additional non-inferential mechanisms to account for "non-propositional" effects might seem overwhelming. But some researchers have been moving away from this narrow conception of inference towards a broader view which allows for a much wider range of inferential procedures, both conscious and unconscious. Sperber and Wilson (2015: 136) comment,

> Not all inferences involve step by logical step derivations of explicit conclusions from explicit premises. Arguably, the vast majority of inferences made by humans and other animals do not involve such derivations.

This broader view of inference is defended in detail in recent work by Hugo Mercier and Dan Sperber (2011, 2017), who sum it up as follows:

> Cognition involves going well beyond the information available to the senses. All that sensory organs get by the way of information, be it in ants or in humans, are changes of energy at thousands or millions of nerve endings. To integrate this information, to identify the events in the environment that have caused these sensory stimulations, to respond in an appropriate manner to these events, cognition must, to a large extent, consist in drawing inferences about the way things are, about what to expect, and about what to do. (Mercier & Sperber 2017: 56)

On this broader view, perception and memory both involve a substantial element of inference, sensorimotor and memory mechanisms are themselves inferential, and utterance comprehension is an inferential process par excellence. If this approach is on the right track, there may be no need to invoke additional non-inferential mechanisms in accounting for "non-propositional" effects.

The output of the comprehension process is an array of propositions taken to constitute the intended import of the utterance. The notion of a proposition, like the notion of an inferential mechanism, can be more or less narrowly construed. For someone who sees propositions as closely related to the sentences of a natural language, the case for treating the output of the comprehension process as consisting not only of propositions but also of "images or states of mind" might again seem overwhelming. From the start, however, relevance theorists have argued for a broader construal on which the gap between propositions and natural-language sentences is much greater than is standardly assumed. As Wilson & Sperber (2012: ix) put it,

> There are always components of a speaker's meaning which her words do not encode... Indeed, we would argue that the idea that

for most, if not all, possible meanings that a speaker might intend to convey, there is a sentence in a natural language which has that exact meaning as its linguistic meaning is quite implausible.

From this perspective, there are many thoughts that cannot be directly encoded into words.

When ad hoc concepts are brought into the picture, it becomes clear that the gap between language and thought exists not only on the level of whole sentences, but also on the level of individual words. Words are standardly seen as corresponding roughly one-to-one to concepts, so that in order to communicate a concept, one merely has to utter the word that encodes it. By contrast, relevance theorists have argued that humans have many more concepts than words, and indeed, that they may create whole swathes of new concepts for very little effort using existing concepts as templates (Sperber 1996; Sperber & Wilson 1998). On this broader construal, ad hoc concepts are capable of capturing fine-grained differences in perception, action or emotion in a way that lexicalised concepts cannot, and are communicated not by encoding them but by providing evidence of one's intention to convey them.

To illustrate, here is an analysis by Kathryn Banks (2018) of what might be communicated by the line *I stood like that, rippling*, from a poem by Mary Oliver describing the experience of watching a heron, poised between earth and sky, shake its wings, and of feeling suddenly at one with nature.[11] The encoded concept RIPPLE denotes (let us say) a variety of undulating movements, some of which will be highly salient in the immediate context: the shaking of a person experiencing intense joy or pain, the shaking of a heron's wings before it explodes into action, the rippling of water, the undulations of light and darkness against the sky. As Banks puts it,

> So readers ... might imagine movements made by the human body in explosive ecstasy; by enormous expansive heron wings interrupting the bird's poise; by undulating waves of water; and perhaps by ripples of light in the sky. The combination of these sensorimotor imaginings gives some indication of what Oliver's experience might be like, of what it might feel like ... to merge with nature.

The ad hoc concept RIPPLING* would be derived by backwards inference from a vast array of implications which render different aspects of this complex expe-

---

[11] The poem, 'Wings', is published in Oliver (1990: 14–15).

rience in a rich and fine-grained way, and would figure in the reader's (provisional) interpretation of the poem, to be adjusted and readjusted in the course of further readings. In this way, sensorimotor mechanisms might contribute both to the explicature and to the vast array of weak implicatures which make up the intended import of the poem, and a single overall framework might handle the derivation of both propositional and "non-propositional" effects.

5 The relevance of fiction

I have tried to show that the "ad hoc concept" account of metaphor has all the advantages of the earlier "no explicature" account while solving a number of problems created by that account. According to the "ad hoc concept" account, loose, hyperbolic and metaphorical utterances express explicatures just as literal utterances do, and are capable of being both relevant and true. There is then little point in pursuing Reboul's suggestion that we might look to metaphor for help in understanding how fictional utterances can be relevant without being true.

Of course, utterances can be relevant in many different ways. As the history of literary theory shows, fiction can be found relevant for many reasons, whether or not one is interested in identifying the author's intentions. However, the question Reboul (1986, 1987) raises is precisely about the intended import of fictional utterances: how can they be relevant *in the way the author intended* without being true?

According to relevance theory, in order to identify the intended import of an utterance, one necessarily has to go beyond it. As we chat at a party, you ask me if I grew up in England and I tell you that I grew up in Cornwall. My reply explicitly communicates that I grew up in Cornwall and strongly implicates that I grew up in England. By implicating that I grew up in England, I specifically answer your question, and therefore have good reason to think that this information will be relevant to you. However, I may have no clear idea *how* it (or the more specific information that I grew up in Cornwall) will be relevant: that is, what implications it will enable you to derive. Yet these are the implications that will satisfy your expectation of relevance. As Sperber and Wilson (1982: 78) put it,

> what our theory of relevance implies is that one of the speaker's intentions (and a crucial one) is that the hearer, by recognising the speaker's intentions, should be made capable of going beyond them and of establishing the relevance of the utterance for himself.

This general intention of being relevant gives the crucial guide to recovery of the meaning, references and inferences (if any) intended by the speaker. A successful act of comprehension (which is what is aimed at by both speaker and hearer) is one which allows the hearer to go beyond comprehension proper.

It may therefore be worth distinguishing *comprehension* from *interpretation,* where comprehension is the process of recognising the intended import of an utterance, and interpretation includes the broader process of drawing one's own conclusions as part of the overall search for relevance. When the intended import consists of a vast array of propositions, there may be no clear cut-off point between comprehension and interpretation. While some propositions in the array may be strongly communicated, others will be more weakly implicated. As communication becomes weaker, comprehension shades off into interpretation, and the borderline between authorial intentions and unintended implications becomes increasingly blurred.

With fictional works, it may also be worth distinguishing two types of relevance, or two ways in which expectations of relevance can arise. Expectations of "internal" relevance arise in the context of the preceding text and guide the interpretation of subsequent text. Typically, the early chapters of a novel help to introduce a context in which later events will achieve relevance of this "internal" type. While some fiction (genre novels, romantic or detective fiction) may have mostly this "internal" type of relevance, literary works are generally expected to have more significance than this. That is, they are typically expected to achieve some "external" relevance by having lasting cognitive effects on beliefs and assumptions about the actual world that the reader has independently of the text. This is the type of relevance that Reboul (2011) has in mind, and it presents a problem for both relevance theory and literary studies. As Reboul notes, works of fiction are not put forward as true descriptions of the actual world, so how can they provide evidence strong enough to achieve lasting effects of this type?[12]

Jerry Fodor (1998: 13) has fun with an attempt by Stephen Pinker to pin down the precise nature of the lasting cognitive effects that fictional works are expected to achieve:

> And here [Pinker] is on why we like to read fiction: "Fictional narratives supply us with a mental catalogue of the fatal conundrums we might face someday and the outcomes of strategies we

---

[12]For a range of different perspectives on the relation between fiction and belief, see Sullivan-Bissett, Bradley & Noordhof (2017).

could deploy in them. What are the options if I were to suspect that my uncle killed my father, took his position, and married my mother?" Good question. Or what if it turns out that, having just used the ring that I got by kidnapping a dwarf to pay off the giants who built me my new castle, I should discover that it is the very ring that I need in order to continue to be immortal and rule the world? It's important to think out the options betimes, because a thing like that could happen to anyone and you can never have too much insurance.

Here is a more promising suggestion from Chomsky (1987: 59):

> It is quite possible—overwhelmingly probable, one might guess—that we will always learn more about human life and personality from novels than from scientific psychology.

This fits well with an idea briefly put forward by Sperber and Wilson (1987: 751), that the author of a work of fiction is simultaneously performing communicative acts on two levels: a lower-level act of describing a fictional world, and a higher-level act of showing this world to the reader as an *example* of what is possible, or conceivable. The expectations of relevance raised by the lower-level act would be internal, while the higher-level act would create external expectations of relevance. Novels in particular provide us with well-described examples of situations, characters and interactions. To the extent that these examples are truth-like, they do not actually have to be true in order for us to learn from them about what is possible or conceivable. Often, the type of relevance they achieve consists in strengthening or reorganising existing assumptions, creating new inferential routines or helping us articulate thoughts of which we had only been dimly aware. Although I have argued that the parallels between metaphor and fiction are not as close as Reboul suggests, this is a type of relevance that many creative metaphors also achieve.

References

Banks K. (2018), "'Look Again', 'Listen, Listen', 'Keep Looking': Emergent properties and sensorimotor imagining in Mary Oliver's poetry", in T. Cave T & D. Wilson (eds). *Reading Beyond the Code: Literature and Relevance Theory*. Oxford: Oxford University Press.

Barsalou L. (1987). "The instability of graded structure: Implications for the nature of concepts". In U. Neisser (ed). *Concepts and Conceptual Development: Ecological and Intellectual Factors in Categorization*. Cambridge: Cambridge University Press. 101–40.

Bezuidenhout A. (2002). "Truth-conditional pragmatics". *Philosophical Perspectives* 16: 105–134.

Bolens G. (2012). *The Style of Gestures: Embodiment and Cognition in Literary Narrative*. Baltimore, MD: Johns Hopkins University Press.

— (2016). *L'Humour et le savoir des corps: Don Quichotte, Tristram Shandy et le rire du lecteur*. Rennes, France: Presses Universitaires de Rennes.

Bradley A.C. (1909). *Oxford Lectures on Poetry*. London: Macmillan.

Carston R. (1997). "Enrichment and loosening: Complementary processes in deriving the proposition expressed?" *Linguistische Berichte* 8: 103–27.

— (2002). *Thoughts and Utterances: The Pragmatics of Explicit Communication*. Oxford: Blackwell.

— (2010). "Metaphor: ad hoc concepts, literal meaning and mental images". *Proceedings of the Aristotelian Society* 110: 295–321.

— (2015). "Lexical pragmatics, ad hoc concepts and metaphor". *Italian Journal of Linguistics* 22: 153–180.

Cave T. (2016). *Thinking with Literature: Towards a Cognitive Criticism*. Oxford: Oxford University Press.

Cave T. & D. Wilson (eds). (2018). *Reading Beyond the Code: Literature and Relevance Theory*. Oxford: Oxford University Press.

Cavell S. (1965). "Aesthetic problems of modern philosophy". In M. Black (ed). *Philosophy in America*. Ithaca, NY: Cornell University Press. Reprinted in S. Cavell (1976). *Must we Mean what we Say?* Cambridge: Cambridge University Press. 73–96.

Chomsky N. (1987). *Language & Problems of Knowledge: The Managua Lectures*. Cambridge, MA: MIT Press.

Fodor, J (1998). "The trouble with psychological Darwinism". *London Review of Books* 20(2): 11–13.

Hauk O., I. Johnsrude & F. Pulvermüller (2004). "Somatotopic representation of action words in human motor and premotor cortex". *Neuron* 41: 301-

307.

Mercier H. & Sperber D. (2011). "Why do humans reason? Arguments for an argumentative theory". *Behavioral and Brain Sciences* 34: 57–111.

— (2017). *The Enigma of Reason,* Cambridge, MA: Harvard University Press, and Harmondsworth: Penguin Books.

Nazir T.A., O. Hauk & M. Jeannerod (2008). "The role of sensory-motor systems for language understanding. Foreword". *Journal of Physiology, Paris* 102: 1–3.

Oliver M. (1990). *House of Light.* Boston, MA: Beacon Press.

Pilkington A. (2000). *Poetic Effects: A Relevance Theory Perspective.* Amsterdam: John Benjamins.

— (2001). "Non-lexicalised concepts and degrees of effability: Poetic thoughts and the attraction of what is not in the dictionary". *Belgian Journal of Linguistics* 15: 1–10.

Reboul A. (1986). "L'interprétation des énoncés de fiction". *Cahiers de linguistique française* 7: 27–41.

— (1987). "The relevance of *Relevance* for fiction". *Behavioral and Brain Sciences* 10: 729.

— (1990). "The logical status of fictional discourse: what Searle's speaker can't say to his hearer". In A. Burkhardt (ed). *Speech Acts, Meaning and Intentions: Critical Approaches to the Philosophy of John R. Searle.* Berlin: de Gruyter. 336–363.

— (1992). *Rhétorique et stylistique de la fiction.* Nancy: Presses Universitaires de Nancy.

— (1997). "La fiction et le mensonge: les 'parasites' dans la théorie des actes de langage". *Psychologie de l'Interaction* 5-6: 87–125.

— (2000). "Communication, fiction et expression de la subjectivité": *Langue française* 128: 9–29.

— (2001). "Represented speech and thought and auctorial irony: ambiguity and metarepresentation in literature". In P. Boggards, J. Rooryck & P. Smith (eds). *Quitte ou double sens: mélanges sur l'ambiguïté offerts à Ronald Landheer.* Amsterdam: Rodopi: 253–77.

— (2008). "L'ironie auctoriale: une approche gricéenne est-elle possible?" *Philosophiques* 35. 25–55.

— (2009). "La fiction, la narration et le développement de la rationalité". *Nouveaux Cahiers de Linguistique française* 29: 83–98.

— (2011). "Live metaphors", in *Philosophical Papers Dedicated to Kevin Mulligan,* Pub. Electronique, Université de Genève, 1–17.

— (2012). "Réception de la fiction: le cas de la résistance imaginative" in B.

Voisin (ed). *Du récepteur ou l'art de déballer son pique-nique*. Rouen, France. 1–7.

Recanati F. (1995). "The alleged priority of literal interpretation". *Cognitive Science* 19: 2007–32.

— (2004). *Literal Meaning*. Cambridge: Cambridge University Press.

Sperber D. (1996). *Explaining Culture: A Naturalistic Approach*. Oxford: Blackwell.

Sperber D. & D. Wilson (1982). "Mutual knowledge and relevance in theories of comprehension". In N. Smith (ed). *Mutual Knowledge*. London: Academic Press. 61–131.

— (1986a). *Relevance: Communication and Cognition*. Oxford: Blackwell. (2nd edition 1995.)

— (1986b). "Loose talk". *Proceedings of the Aristotelian Society* 86: 153-71. Reprinted in S. Davis (ed.), (1991). *Pragmatics: A Reader*. Oxford: Oxford University Press. 540–49.

— (1987). "Presumptions of relevance". *Behavioral and Brain Sciences* 10: 736–54.

— (1998). "The mapping between the mental and the public lexicon". In P. Carruthers & J. Boucher (eds). *Language and Thought: Interdisciplinary Themes*. Cambridge: Cambridge University Press. 184–200. Reprinted in Wilson & Sperber (2012): 31–46.

— (2008). "A deflationary account of metaphors". In R. W. Gibbs (ed). *Metaphor in Language and Thought*. Cambridge: Cambridge University Press. 84–105. Reprinted in Wilson & Sperber (2012): 97–122.

— (2015). "Beyond speaker's meaning". *Croatian Journal of Philosophy* 15(44): 117–49.

Sullivan-Bissett E., H. Bradley & P. Noordhof (eds). (2017). *Art and Belief*. Oxford: Oxford University Press.

Vega Moreno R. (2007). *Creativity and Convention: The Pragmatics of Everyday Figurative Speech*. Amsterdam: John Benjamins.

Wilson D. & R. Carston (2006). "Metaphor, relevance and the 'emergent property' issue". *Mind & Language* 21: 404–33.

— (2007). "A unitary approach to lexical pragmatics: Relevance, inference and ad hoc concepts". In N. Burton-Roberts (ed.). *Pragmatics*. Basingstoke: Palgrave. 230–59.

Wilson D. & P. Kolaiti (2017). "Lexical pragmatics and implicit communication". In P. Cap & M. Dynel (eds). *Impliciteness: From Lexis to Discourse*. Amsterdam: John Benjamins. 147–175

Wilson D. & D. Sperber (1981). "On Grice's theory of conversation". In

P. Werth (ed). *Conversation and Discourse*. London: Croom Helm. 155–178.
— (2002). "Truthfulness and relevance". *Mind* 111: 583–632. Reprinted in Wilson & Sperber (2012): 47–83.
— (2012). *Meaning and Relevance.* Cambridge: Cambridge University Press.

Part II

Language evolution

Chapter 9

Language Evolution: Insisting on Making It a Mystery or Turning It into a Problem?

Pedro Tiago Martins & Cedric Boeckx

Abstract Hauser et al. (2014) offer a rather negative view of the state of affairs in language evolution. The authors target some of the fields that have spawned the most activity and hypotheses in recent years (comparative animal behavior studies, archaeology, molecular biology and modelling), and then show what they have done wrong. We argue instead that it is the progress in these fields that accounts in large part for the revival of biolinguistic concerns.

Keywords Evolution of language · Biolinguistics · Comparative animal behavior · Archaeology · Molecular biology · Evolutionary modelling.

In a recent, widely-read paper, Hauser et al. (2014) offer a rather negative view of the state of affairs in language evolution. More specifically, the authors believe that little to no progress has been made in the various relevant fields regarding the age-old questions of the origin and evolution of the human capacity for language. We beg to differ.

The authors' strategy is to target some of the fields that have spawned the most activity and hypotheses in recent years (comparative animal behavior studies, achaeology, molecular biology and modelling), and then show what they have done wrong. These fields, they say, have not advanced much more than speculation. Instead, we think that it is the progress in these fields that accounts in large part for the revival of biolinguistic concerns (Boeckx, 2013).

The intention of Hauser et al.'s paper is to point out the damage that has been done over the last decades, by calling attention to the dangers of jumping from simplistic, impoverished data and observations to full-fledged accounts. To some extent, we agree. But we find it curious that linguistics is not of the targets of the paper, even though the field is rife with speculative and untestable proposals and implications for how language evolved. The implicit but in our view obvious corollary is that—for the authors—linguistic theorizing plays at present a crucial role in advancing what we know about language evolution, or at the very least does not have much to be criticized (while other fields do have a lot to be criticized for, since they do not match what has been or could be accomplished by linguistic theorizing). We take this absence with a grain of salt, as we find it hard to explain how a paper on the status of language evolution studies does not even dabble in the shortcomings of what is in effect the field of expertise of half of its 8 co-authors.

While it is true that we do not know how language evolved—if we did, no one would be working on it anymore—, to diminish the work in that has been done recently on various disciplines to the point of irrelevancy is not only dubious (we feel it ends up throwing the baby with the bathwater) but, in the case of this group of authors, confusing (a close look at the literature will reveal that different combinations of the authors of Hauser et al. (2014) make arguments of the sort they take issue with, and rely on sources of information that in the paper are deemed unreliable). In what follows we will briefly touch on the different fields targeted by Hauser et al. (2014), and point out both incongruence and unjustified pessimism in their arguments. We will not offer here in-depth rebuttals or qualifications of the authors' positions, but instead provide a little glimpse into what we see as more heat than light.

In relation to the archaeological record, which in the abstract the authors say "does not inform our understanding of the computations and representations of our earliest ancestors, leaving details of origin and selective pressure unresolved", Chomsky (2005, 3), on the basis of work by Tattersall, writes of the faculty of language as part of a "a complex of capacities that seem to have crystallized fairly recently, perhaps a little over 50,000 years ago, among a small breeding group of which we are all descendants—a complex that sets humans apart rather sharply from other animals, including other hominids, judging by traces they have left in the archaeological record." In the very same page, Chomsky goes on to say that the great leap forward is "the result of some genetic event that rewired the brain, allowing for the origin of modern language with the rich syntax that provides a multitude of modes of expression of thought, a prerequisite for social development and the sharp changes of behavior that are revealed in the archaeological record [...]." The same ideas are echoed, for example, in Chomsky (2010). Yang (2010) claims that we cannot ask too much of Universal Grammar, because "[a] theory of Universal Grammar is a statement of human biology, and one needs to be mindful of the limited structural modification that would have been plausible under the extremely brief history of *Homo sapiens* evolution." But how do we know this if, according to the same authors, language evolution has been a complete mystery for years? Speculation goes both ways, and one should not dismiss one and support the other. It seems that for Hauser et al. (2014) arguments of this sort were appropriate while the relevant archaeological record was thought to have been left by humans, and only now that this is not so clear (e.g. Zilhão, 2011), the authors claim we shouldn't try to derive inferences from archaeology. Nonetheless, one needs not look hard to find resort to archaeological evidence in support of a non-gradualist position as recently as in 2016 (e.g. Berwick and Chomsky, 2016, 37-38), leaving us all the more confused as to what their overall position regarding its reliability as a source of information really is.

Hauser et al. (2014) also take issue with comparative animal work, which "provide[s] virtually no relevant parallels to human linguistic communication, and none of the underlying biological capacity." The problem with this assessment is that it equates the testing of all-or-nothing hypotheses (animal X displays some form of language phenotype property P) with everything that such endeavors might have to offer. We take it that not many people believe in "talking birds" and "signing apes" (if this was ever the case for serious scientists), but those studies and their scrutiny are important to determine what humans and non-human animals do, and the wealth of animal studies is becoming increasingly more important in the study of underlying mechanisms shared by different species and formulation of hypothesis concerning humans in particular. Berwick and Chomsky (2013) seem like they would agree, and Berwick et al. (2011), for example, draw a connection between birdsong syntax and underlying mechanisms of human speech, and state that "comparing the structure of human speech and birdsong can be a useful tool for the study of evolution of brain and behavior" (p. 120). This qualifies as Hauser et al's (2002) FLN, which in the present paper the authors stress as referring not only to the mechanisms for discrete infinity but also to the "mappings to the interfaces with the conceptual-intentional and sensory-motor systems." Hauser et al. (2014) are right to point out that some current techniques used in animal studies fail to capture the animals' actual capacities, which they are more likely to display in their natural habitats, roaming free and devoid of extensive, goal-oriented training. But by doing so they are targeting the search for a human-like linguistic phenotype in other species, rather than the bottom-up comparative work of the kind advocated, for example, by de Waal and Ferrari (2010), to which we will return later.

As for molecular biology, Hauser et al. (2014) do not present a critique *per se*, but rather an overview of current work which shows that there is no clear path from genes to linguistic behavior. This is not surprising to molecular biologists, and in fact simplistic proposals of the sort they criticize—coming up with just-so stories out of thin air, or on the basis of impoverished observations—usually come from the field of linguistics.[1] It is for this reason that work in linguistics must provide information that can be used to creating linking hypotheses, which currently and for the most part it cannot. This difficulty in creating linking hypothesis between genes and linguistic behavior is amplified by this inadequacy of linguistics in providing primitives that other fields can work with.[2] A logical theory of the language faculty does not necessarily amount to a biologically plausible one, which is what we should be aiming for. This state of affairs alone would warrant a discussion of linguistics as a source of information in language evolution studies that is absent from Hauser et al. (2014). The way in which the authors present the linguistic phenotype—a novel recursion mechanism, unique to humans—is enough to stall or severely hinder the kind of linking hypotheses we all would like to see, and which Hauser et al. (2014) say we have no hopes of seeing any time soon. The reason for that is that we actually know that novelty doesn't simply "arise". While traits may on the surface seem novel, or *sui generis*,[3] their nature is "largely reorganizational, rather than the product of innovative genes" (West-Eberhard, 2005, p. 6547), that is, phenotypic novelties are the result of the combination of different, more generic mechanisms.

Hauser et al.'s case against current work in modelling is the most consistent with each author's practice, but their general disdain for the role of culture in evolution—"In this paper, we are interested in biological *as opposed* to cultural evolution" (p. 2, our emphasis)—overlooks important advancements in evolutionary biology which show that culture *and* environment might really be crucial. "Culture" is a taboo notion in most generative circles, perhaps because it is usually seen as detrimental to biology in a theory of language. We find this to

---

[1] See Boeckx (2016) for discussion of a recent example.
[2] For a discussion of this problem, see Poeppel and Embick (2005).
[3] For discussion, see Wagner and Müller (2002) and Moczek (2008), among others.

happen only under a naïve view of biology, along with an axiomatic incompatibility with linguistic approaches that give pride of place to culture. Crucially, one should not ignore the role of environmental factors in the shaping of the genotype, and in turn the shaping of the phenotype. There is no reason to seek explanation of phenotypic variation only in environmental or genetic factors. Instead, one should incorporate the lessons from Evo-Devo, and pay attention to work on the genotype-environment interaction (West-Eberhard, 2003), which shows that the degree to which environmental choices affect the way the genetic blueprint is expressed depends on the specific genotype-environment interaction in each case.

In a somewhat more optimistic tone, Hauser et al. (2014) offer some suggestions of "paths forward", both interspersed throughout the paper and as a final comment. These suggestions, however, are very much confined, and suffer from the same problems that their negative assessment of the various fields does. In a nutshell, the authors insist on gauging the usefulness of theoretical and experimental work by whether or not it "speaks" to Merge, the recursive mechanism they place at the center of the linguistic phenotype. It is not surprising that the presupposition that Merge must be at the center of inquiry into language evolution drastically reduces what can be done in practice, but in doing so it pushes the *mystery* the authors speak of. That is not what parsimony is for. Language evolution thus becomes a *mystery* only to adherents of this presupposition, and a *problem*—like many others in the sciences—for those willing to explore further.

In the case of animal studies, the authors put their money on the development of new techniques that could allow the collection of neural data from free-ranging animals, thus revealing their capacities in the absence of reinforcement. We agree that such techniques would work wonders for the field, but what propels Hauser et al. (2014) is that we would then be able to devise a "set of stimuli that are generated from a recursive operation such as Merge (a recursive operation that combines two objects, such as two lexical items, to construct a new object, such as a phrase, in a process that can be iterated indefinitely), expose animals to a subset of these, and then test them on a wide range of alternatives that extend beyond the initial set in ways that can reveal substantial generalization, and thus comprehension of the underlying generative operation." (p. 9–10) Presumably, these tests would reveal that animals either fail miserably or are able to generalize by relying on different, finite mechanisms, thus showing the uniqueness of Merge and supporting the discontinuity hypothesis. But there are myriad (other) ways in which animal studies can work in favor of a deeper knowledge about the biology of language. In this context, we find it appropriate to quote a passage by (de Waal and Ferrari 2010: 201):

> Over the last few decades, comparative cognitive research has focused on the pinnacles of mental evolution, asking all-or-nothing questions such as which animals (if any) possess a theory of mind, culture, linguistic abilities, future planning, and so on. Research programs adopting this top-down perspective have often pitted one taxon against another, resulting in sharp dividing lines. Insight into the underlying mechanisms has lagged behind. A dramatic change in focus now seems to be under way, however, with increased appreciation that the basic building blocks of cognition might be shared across a wide range of species. We argue that this bottom-up perspective, which focuses on the constituent capacities underlying larger cognitive phenomena, is more in line with both neuroscience and evolutionary biology.

Indeed, looking for a full-fledged ability such as language or something that looks close enough to it is bound not to tell us much, but that's not what we should be looking for. Instead, we should decompose it into more generic mechanisms, not unique to either the language domain nor the human species. This path will inevitably leads us to the study of abilities with little resemblance to language, and mechanisms at levels far deeper than the behavioral and the cognitive. But it's these levels we need to get to in order to arrive at true linking hypotheses.

As for modeling, the authors say that "it must focus on the computations and representations of the core competence for language, recognize the distinction between these internal processes and their potential externalization in communication, and lay out models that can be empirically tested in our own and other species." Again, it must speak to Merge (which is how we must interpret "the core competence for language" when reading Hauser et al. (2014)), and a host of other possible modeling work is not even considered. We don't see how this would change the status of the field if all we are allowed to focus on is the core recursive mechanisms the authors equate with the linguistic phenotype (and perhaps the interfaces between and externalization systems, which are usually left vague in any case). Opening one's mind to the role of the environment (or culture, which we find hard to tease apart from "environment" in a meaningful way) is likely to prove fruitful, and modeling work pays particular attention to the influence this aspect of the world might have. We agree with (Kirby, 2013, 473) that "the particular learning mechanisms that we bring to bear on the task of acquiring language are assuredly part of our biology. The key questions to ask about this aspect of the biology of language are: what is the nature of our biological endowment that relates to language learning? and to what extent is this endowment specific to language? These questions essentially define the biolinguistics enterprise, and their answer depends on an understanding of the relationships between learning, cultural transmission, and biological evolution."

In sum, Hauser et al. (2014) paint an ugly picture of language evolution that seems to have been caused by other, incautious scientists, while in reality the authors themselves have incurred in the same kind of arguments and assumptions—the kind they deem poor and speculative. This practice has not stopped with this paper: a quick read through the latest book by two of the authors (Berwick and Chomsky, 2016) will reveal discussion of topics that in Hauser et al. (2014) we are advised not to pay much attention to. This kind of incongruous back-and-forth is bound to cause more confusion than resolution. Furthermore, insisting on the idea that the evolution of language is mysterious—and not a problem we can look into right now, with its own difficulties and promising avenues—will deter only those who are stuck with a naïve view of biology that allows for such a simplistic perspective.

What is clear to us, and not so clear from reading Hauser et al. (2014), is that in order to make language evolution more of a problem and less of a mystery, everyone—linguists included—will have to make the mapping between mind to brain the focus of study. It is this intermediate level between genotype and phenotype that must be the target of intensive investigation. If the mind is what the brain does, it is imperative to understand how the brain came to do what it does. This will necessarily involve a reconsideration of the nature and fabric of the language faculty, for only those descriptions of linguistic knowledge that can be associated with concrete neural correlates will have a fighting chance of going beyond the limitations of the fossil record, and exploit findings in paleoneurology, paleogenetics, and comparative cognitive biology.

References

Berwick R.C. & N. Chomsky (2013). "Foreword: A bird's-eye view on human language and evolution". In J. J. Bolhuis and M. Everaert (eds). *Birdsong, Speech, and Language*. Cambridge, MA: MIT Press. ix–xii.
— (2016). *Why only us. Language and Evolution*. Cambridge, MA: MIT Press.
Berwick R.C., K. Okanoya, G.J. Beckers & J.J. Bolhuis (2011). "Songs to

syntax: the linguistics of birdsong". *Trends in cognitive sciences* 15(3): 113–121.
Boeckx C. (2013). "Biolinguistics: Forays into human cognitive biology". *Journal of Antropological Sciences* 91: 63–89.
— (2016). "Ljiljana Progovac, *Evolutionary syntax*". *Journal of Linguistics* 52(2): 476–480.
Chomsky N. (2005). "Three factors in language design". *Linguistic inquiry* 36(1): 1–22.
— (2010). "Some simple evo devo theses: How true might they be for language". In R. K. Larson, V. Déprez and H. Yamakido (eds). *The Evolution of Language: Biolinguistic Perspectives*. Cambridge: Cambridge University Press. 45–62.
de Waal F. & P.F. Ferrari (2010). "Towards a bottom-up perspective on animal and human cognition". *Trends in cognitive sciences* 14(5): 201–207.
Hauser M.D., N. Chomsky & W.T. Fitch (2002). "The Faculty of Language: What is It, Who has It and How Did It Evolve?". *Science* 298: 1569–1579.
Hauser M.D., C. Yang, R.C. Berwick, I. Tattersall, M. Ryan, J. Watumull, N. Chomsky & R. Lewontin (2014). "The mystery of language evolution". *Frontiers in Psychology* 5(401): 401.
Kirby S. (2013). "Language, culture, and computation: An adaptive systems approach to biolinguistics". In C. Boeckx and K. K. Grohmann (eds). *The Cambridge Handbook of Biolinguistics*. Cambridge: Cambridge University Press. 460–477.
Moczek A.P. (2008). "On the origins of novelty in development and evolution". *BioEssays* 30(5): 432–447.
Poeppel D. & D. Embick (2005). "Defining the relation between linguistics and neuroscience". In A. Cutler (ed). *Twenty-First Century Psycholinguistics*. Mahwah NG, London: Lawrence Erlbaum. 103–118.
Wagner G.P. & G.B. Müller (2002). "Evolutionary innovations overcome ancestral constraints: a re-examination of character evolution in male sepsid flies". *Evolution & Development* 4(1): 1–6.
West-Eberhard M.J. (2003). *Developmental plasticity and evolution*. Oxford: Oxford University Press.
— (2005). "Developmental plasticity and the origin of species differences". *Proceedings of the National Academy of Sciences of the United States of America*, 102(Suppl 1): 6543–6549.
Yang C. (2010). "Three factors in language acquisition". *Lingua* 120: 1160–1777.

Zilhão J. (2011). "The emergence of language. art and symbolic thinking: A neandertal test of competing hypotheses". In C. S. Henshilwood and F. d'Errico (eds). *Homo Symbolicus. The Dawn of language, imagination and spirituality*. Amsterdam: Johns Benjamins. 111–131.

Chapter 10
On Language not Being at Root a Communication System. Some Morphosyntactic Considerations

Frederick J. Newmeyer

Abstract Anne Reboul has argued that enhanced communicative abilities could not have been the driving force in the origin and biological evolution of language. This paper supports Reboul's hypothesis from the point of view of morphosyntax. It provides arguments to reject the incrementalist approach to language evolution proposed by Ray Jackendoff and others, in which the evolution of morphosyntax has proceeded in biologically-determined stages, each stage improving overall communicative efficiency.

Keywords Evolution of language · Communication · Morphosyntax

1 Introduction

This paper is a follow up to Anne Reboul's article 'Why language really is not a communication system: A cognitive view of language evolution' (Reboul 2015). Reboul develops several convincing arguments in support of the idea that enhanced communicative abilities could not have been the driving force in the origin and biological evolution of language. As she notes, the core features of language — semanticity, discrete infinity, and decoupling — are not found in other known systems of communication, rendering implausible the idea that language evolved for purposes of communication. Reboul goes on to demonstrate that two prominent models of communication systems — the code model (Millikan 2005) and the ostensive model (Scott-Phillips 2015) — cannot account for language evolution. However, Reboul's article has little to say about the evolution of the grammatical properties of language, in particular morphology and syntax. My goal here is to argue that these properties did not arise evolutionarily to support communication. While, to be sure, morphosyntax is in a certain sense 'designed' for communication, this design was effectuated over historical time, rather than evolutionary time.

Reboul's hypothesis that language arose as a cognitive, as opposed to a communicative, tool was at one time the standard view of formal linguists. By 'formal linguists' I mean those working in the tradition initiated by Noam Chomsky, which accords center stage to the structural properties of grammatical phenomena, and characterizes these properties by means of an algebraic system called a 'generative grammar'. Here are some quotes from representative work, with key passages emphasized:

> We should search for the ancestry of language not in prior systems of animal communication but *in prior representational systems* (Bickerton 1990: 23)

> A far better case could be made that grammar exploited mechanisms *originally used* for the conceptualization of topology and antagonistic forces [than for motor control] (Jackendoff 1983, Pinker 1989, Talmy 1983, Talmy 1988), but that is another story. (Pinker & Bloom 1990: 726)

> The syntactic category system and the conceptual category system match up fairly well. In a way, the relation between the two systems serves as a partial explication of the categorial and functional properties of syntax: *syntax presumably evolved as a means to express conceptual structure*, so it is natural to expect that some of the structural properties of concepts would be mirrored in the organization of syntax. (Jackendoff 1990: 27)

> The conditions for the *subsequent development* of language as a medium of communication were set by the evolution of ... the level of conceptual structure ... A *first step* toward the evolution of this system for communication was undoubtedly the linking up

of individual bits of conceptual structure to individual vocalizations ... (Newmeyer 1991: 10)

[T]he emergent ability, driven by the evolutionary appearance of C[onceptual] S[tructure], was the capacity to acquire meaningful, symbolic, abstract units ... *it would be appropriate to expect adaptation-based explanations to come into play at a later stage, once language came to be used preferentially as the human communication system.* (Wilkins & Wakefield 1995: 179)

As functions are usually informally defined, then, it doesn't make much sense to say that the function of language is communication. ... So some small genetic change led to the rewiring of the brain that made this human [linguistic] capacity available. ... *And most of it is thinking and planning and interpreting, and so on; it's internal.* (Chomsky 2012: 12-14)

As Reboul notes, the bulk of the body of work on language evolution does not share this perspective it all. Rather, it looks at language as a cultural tool for communication and tries to localize its origins in some (predominantly) cultural change, whose result is increased communicative success. Among the hypothesized triggering factors are shared intentionality (Tomasello, Carpenter, Call, Behne, & Moll 2005); a bonding mechanism in order to use social time more efficiently (Aiello & Dunbar 1993); social grooming (Dunbar 1996); female coalitionary strategies (Power 1998); female choice of mate (Miller, 2001); territorial scavenging (Bickerton 2009); communicative strategies involved in hunting (Washburn & Lancaster 1968; Hewes, 1973); need for mother-child communication (Falk 2004); and foraging efficiency among hunter-gatherers (MacDonald & Roebroeks 2013).

There is certainly nothing amiss about looking at cultural prerequisites for communication. But a cultural focus is not likely to lead to an understanding of why grammatical systems have the architectural properties that they have, that is, what determines the set of grammatical categories and relations across languages, why rules, principles, and constraints are remarkably similar from language to language, why long-distance dependencies exist and how they are constrained, why some word orders are more common than others, and how syntax, morphology, semantics, and phonology interact.

The question, then, is how to account for the origins of these broad design features of morphosyntax. The prevailing opinion among generative grammarians since the 1960s has been that the grammatical system is not only situated in the human mind, but also that its fundamental principles, its inventory of combinatorial elements, and so on are innate (see Chomsky 1965, Chomsky 1981). An innate property is, by definition, encoded in the genome, which immediately raises the question of what sort of evolutionary event(s) could have engendered a rich innate component to morphosyntax. Until the last decade or so, generative grammarians have had no answer to this question, except to reaffirm that the enhancement of communicative skills had nothing to do with it. But recently, some prominent generativists have reconsidered the issue and have come to conclude that important features of morphosyntax were selected for evolutionarily on the basis of their enhancement of communicative abilities. The following §2 sketches two of these approaches and discusses briefly how Chomsky's most recent work is not as 'anti-communication' as one might conclude from a casual reading of some of his publications. Section 3 presents a critique of the idea that the design properties of morphosyntax arose over evolutionary time by virtue of their leading to enhanced communicative abilities. Section 4 is a brief conclusion.

2 Some communication-oriented generative models of the evolution of morphosyntax

The following subsections discuss three grammatically-informed models of language evolution: Ray Jackendoff's (§2.1), Ljiljana Progovac's (§2.2), and Noam Chomsky's (§2.3).

2.1 Ray Jackendoff's communicatively-oriented model

Ray Jackendoff has been a central figure of formal linguistics for a half century. The defence and refinement of the theory of UG has always been one of his major concerns, as is indicated by the following representative quote:

> Lines of evidence from the structure of numerous languages, from historical changes in languages, from the character of child language acquisition, and from linguistic deficits due to brain damage all converge on the view that there is a highly specified innate bias ('Universal Grammar') from which children develop an adult language capacity during the first ten or twelve years of life. This innate bias ... seems to be a specific brain adaptation, specialized to deal with this particular eccentric form of information we call language. (Jackendoff 1992: 71)

Furthermore, Jackendoff has always assumed the correctness of the autonomy of syntax (for a lucid defence of autonomy, see Jackendoff 1990: 285-286). His current theory of grammar, 'Simpler Syntax' (Culicover & Jackendoff 2005) differs markedly in many respects from Chomsky's approach, though these differences and, indeed, the details of the model, need not concern us in this paper.

What distinguishes Jackendoff's approach to language evolution from all other approaches is its extreme incrementalism. In this model, language evolution has proceeded in stages, each stage improving communicative efficiency. Fig. 10.1 illustrates.

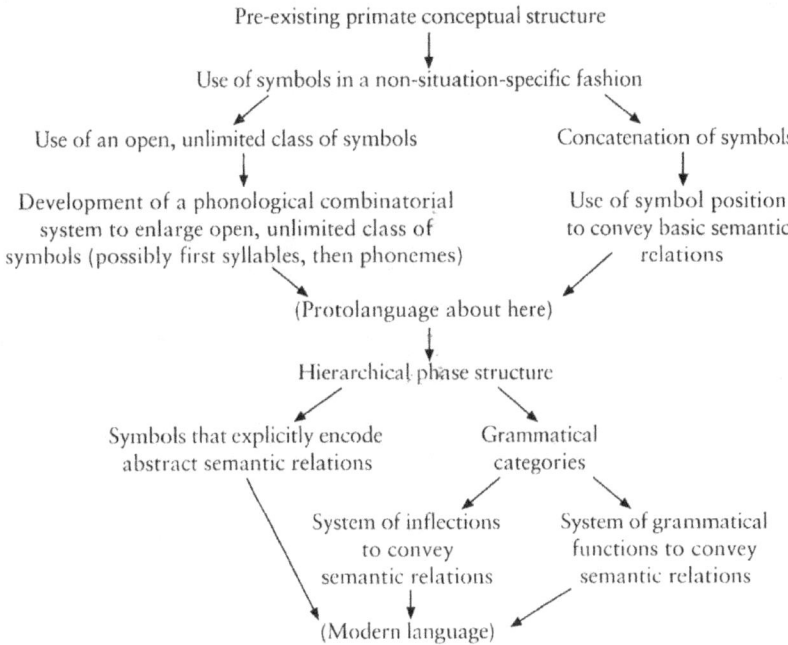

Figure 10.1: The incremental evolutionary steps posited in Jackendoff (2002)

Each stage in the progression represents a change to the human genome, shaped either by a mutation or by the Baldwin effect: 'Language has evolved incrementally in response to natural selection' (Jackendoff 2002: 235). To give a specific example, he has noted that 'one can see the selective advantage in adding grammatical devices that make overt the intended relationships among words — things like word order and case marking'. (Jackendoff 1997: 17). The same can be said about the system of grammatical functions to convey semantic relations (the GF-tier):

> Thus we are pressed into the position of claiming that there is something innate about [the GF-tier] — it is one of those things that has to be carried on the genome. ... Our conjecture is that it is a late evolutionary add-on to the syntax-semantics interface, hardly inevitable but affording adaptive advantages in efficiency and reliability of communication ... (Culicover & Jackendoff 2005: 539)

In Jackendoff's theory, the earlier stages of evolution are still visible as 'living fossils' in certain speech varieties. So Protolanguage manifests itself today in the agrammatic speech of Broca's aphasics, in pidgins, and in the 'Basic Variety' spoken by second language learners. By means of 'reverse engineering' we can use these forms of speech to help reconstruct ancestral versions of human language:

> [H]ow are we to study the evolution of language? To me, the most productive methodology seems to be to engage in reverse engineering. We attempt to infer the nature of universal grammar and the language acquisition device from the structure of the modern language capacity, using not only evidence from normal language, but also evidence from language deficits, language acquisition, pidgins/creoles (Bickerton 1981, DeGraff 1999), and language creation à la [Nicaraguan Sign Language] (Kegl, Senghas, & Coppola 1999) and Israeli Bedouin Sign Language (Sandler, Meir, Padden, & Aronoff 2005). This is what linguists and psycholinguists normally do. (Jackendoff 2010: 65)

Finally, in this approach, UG is a 'toolkit', in the sense that not all languages need choose all of its elements:

> Universal Grammar is not supposed to be what is universal among languages: it is supposed to be the 'toolkit' that a human child brings to learning any of the languages of the world. ... When you have a toolkit, you are not obligated to use every tool for every job. Thus we might expect that not every grammatical mechanism provided by Universal Grammar appears in every language. (Jackendoff 2002: 75)

Given the toolkit approach, one would therefore not be disconcerted to discover a language lacking phrase structure or recursion or the GF-tier or some other UG-provided feature.

2.2 Ljiljana Progovac's communicatively-oriented model

Progovac (2015) assumes the basic picture of minimalist clause structure, as illustrated in (1), where a VP (or small clause) layer is embedded in a TP layer, which is itself embedded in a CP layer:

(1)
$$\begin{array}{c} \text{CP layer} \\ \triangle \\ \text{TP layer} \\ \triangle \\ \text{VP / small clause layer} \\ \triangle \end{array}$$

It is postulated that each higher level was added gradually over evolutionary time. In Progovac's view, the addition of each new level of structure was functionally motivated, in that the human communicative capacity was thereby increased:

> For example, each step in the progression from one-word stage (no syntax), to small clause stage (paratactic two-slot syntax), to hierarchical TP stage accrues clear incremental communicative benefits. Small clauses (or half-clauses), with only one layer of structure, would have been immensely useful to our ancestors when they first started using syntax. A half-clause is still useful, even in expressing propositional content — much more useful than having no syntax at all (one-word stage), and much less useful than having more articulated hierarchical syntax of the specific functional category stage. This is exactly the scenario upon which evolution/selection can operate. (Progovac 2015: 15)

2.3 Chomsky and the evolution of morphosyntax

The most visible UG-based approach to grammar (and the evolutionary origins of grammar) is the Minimalist Program (MP), first elaborated in (Chomsky 1995). For Chomsky, the evolutionary 'great leap forward' (Chomsky 2012: 14) was the mutation enabling the Merge operation, that is, the mutation that allowed unbounded recursivity in language. Chomsky is clear that this event was of great *cognitive* functionality: 'As soon as you have [Merge], you have an infinite variety of hierarchically structured expressions [and thoughts] available to you' (p. 14). One assumes that Chomsky's position is that it was only later in the course of human history that recursivity was exploited for its communicative benefits. For to posit a second mutation enabling the externalization of Merge for use in discourse would amount to a concession that language did arise — in part — to serve the needs of communication. Frankly, I am skeptical that the intricate coordination of mental Merge with the input and output channels used for communication could have been achieved over historical, as opposed to evolutionary, time.

What is more, nobody — and least of all Chomsky — questions that there is more to grammar *as a whole* than recursion. A sampling of the minimalist literature in recent years reveals principles governing morphology, phonology, and formal semantics; agreement, labeling, transfer, probes, goals, deletion; economy principles such as Last Resort, Relativized Minimality (or Minimize Chain Links), and Anti-Locality (which don't fall out from recursion per se, but rather represent conditions that need to be imposed on it); the entire set of mechanisms pertaining to phases, including what nodes count for phasehood and the various conditions that need to be imposed on their functioning, like the Phase Impenetrability Condition; the categorial inventory (lexical and functional), as well as the formal features they manifest; the set of parameters (hundreds have been proposed), their possible settings, and the implicational relations among them. To the extent that these principles are built into UG (and we have no reason to think that minimalists believe otherwise), then they demand an evolutionary explanation. Could any of them have arisen to facilitate communication? The answer is 'Possibly'.

One needs also to refer to Chomsky's discussion of what he calls 'the three factors in language design', namely, genetic endowment, experience, and principles not specific to the faculty of language (Chomsky 2005). Having stressed the centrality of recursion and minimizing (as always) the role of experience, Chomsky appeals heavily to third factor principles:

> The MP seeks to approach the problem 'from the bottom up': How little can be attributed to UG while still accounting for the variety of I-languages attained, relying on third factor principles. (Chomsky 2007: 4)

As far as third factor principles are concerned:

> [they fall] into several subtypes: (a) principles of data analysis that might be used in language acquisition and other domains; (b) principles of structural architecture and developmental constraints that enter into canalization, organic form, and action over a wide range, including principles of efficient computation, which would

> be expected to be of particular significance for computational systems such as language. It is the second of these subcategories that should be of particular significance in determining the nature of attainable languages. (Chomsky 2005: 6)

'Principles of data analysis' and wide ranging 'developmental constraints' are subject to a wide variety of interpretations. As has been frequently noted,[1] parsing principles—that is principles that aid communication by facilitating speech production and comprehension—fit comfortably into the domain of third-factor explanations.

As an example of how purely formal principles have been replaced by communication-based ones within the general envelope of the MP, consider the following. Kayne (1994) provided an elaborate UG-based parametric explanation of why rightward movement is so restricted in language after language. Ackema & Neeleman (2002) on the other hand, argue that the lack of acceptability of structures in 'right-displaced' position should not be accounted for by syntax proper (that is, by the theory of competence), but rather by the theory of language processing. In other words, they provide a communication-based third-factor explanation.

It is also worth pointing out that the evolutionary appearance of grammar, given minimalist assumptions, was not necessarily abrupt, despite the claims in the following quote:

> There is no possibility of an 'intermediate' syntax between a non-combinatorial syntax and full natural-language syntax — one either has Merge in all its generative glory, or one has effectively no combinatorial syntax at all, but rather whatever one sees in the case of agrammatic aphasics: alternative cognitive strategies for assigning thematic roles to word strings. (Berwick 1997: 248)

But Brady Clark has pointed out the MP does allow for gradualism in language evolution:

> In conclusion, syntactocentric architectures like the one presupposed by most work within the Minimalist Program are compatible with an incremental view of the evolution of syntax. The evolution of syntax on minimalist assumptions must have involved several distinct stages, including the evolution of Merge, the evolution of words, and externalization. One or more of these stages

---

[1] See the comprehensive discussion in (Mobbs 2014).

(for example, the emergence of Merge) might have involved further stages, once FLB and FLN are distinguished. (Clark 2013: 191)

Once again, any of the steps of the gradual progression might have been incorporated into the genome on the basis of its facilitation of communication. In sum, despite what one might conclude from some passages in the literature, Chomsky's Minimalist Program does indeed allow for language evolution to have been shaped in part by the needs of communication.

3 Communicative needs did not shape the biological evolution of morphosyntax

This section closely examines the incrementalist approach to the evolution of grammar and concludes that it lacks convincing empirical evidence. In short, there is very little evidence that all of the stages represented in Fig. 10.1 arose in the course of biological (as opposed to historical) evolution. The section addresses the following three questions, each of which is discussed in a separate subsection: Do the incremental steps aid communication to the point that they would be incorporated into the genome? (§3.1) Is 'decremental' development attested? If so, how can that be reconciled with an incremental model? (§3.2) Are there alternative explanations for the development of morphosyntactic complexity that do not involve biological evolution? (§3.3).

3.1 Do the incremental steps aid communication to the point that they would be incorporated into the genome?

It is clear that the earlier (pre-Protolanguage) steps of the Jackendoff progression are selectionally advantageous. The ability to use an unlimited class of symbols and to concatenate them was obviously a great leap forward evolutionarily. There is selective advantage in being able to convey an unlimited range of meanings in a communicative setting, and in particular to be able to exchange information about events distant in time and space. But what about the later steps in the progression: the steps that added grammatical categories, grammatical relations, and so on? That is not clear. Jackendoff asserts that they were indeed incorporated into the genome, in that he 'agree[s] with practically everyone that the "Baldwin effect" had something to do with [how increased expressive power came to spread through a population]' (Jackendoff 2002: 237).

Was the gain in expressive power (if it was indeed accomplished) sufficient for the Baldwin effect to do its work? It is true that a more elaborated grammar can overtly express more semantic nuances, but at the same time elaboration can slow down production (if a lot needs to be made overt).[2] Grammaticalization-related changes can offset this slowdown to a certain extent, by shortening the time needed to express certain frequently-called upon concepts (affixes are faster to produce than words). Viewed from another angle, a less elaborated grammar seems to place more interpretive demands on the hearer, but requires less syntactic processing. We are still at the early stages of being able to measure whether overall communication is enhanced by more elaboration.[3]

The strongest reason to doubt that biological evolution has shaped grammatical elaboration is the fact that there are languages that do very well without some of the putatively evolved devices. So there are languages that present little or no evidence of hierarchical phrase structure, such as the Australian languages Warlpiri, Jiwarli, and Wambaya (Hale 1983, Nordlinger 1998, Austin 2001). There are languages (see Van Valin & LaPolla 1997, Kibrik 1997) that present no evidence for grammatical relations. Based on the work of Mark Durie (1985, 1987), Van Valin and LaPolla identify Acehnese as one such language and Primus (1999) makes the same point with respect to Guarani and Tlingit. No learner of these languages would be led to posit a distinction between subjects and objects. There are languages lacking sentential recursion, or where, at least, recursion is highly restricted (Pirahã; see Everett 2005). And there are languages lacking major categorial distinctions (Riau Indonesian; see Gil 2007). The question is whether these languages are less expressive than languages with the 'full set' of features. Gil says such is not the case for Riau (see especially Gil 2009). In any event, how would one test the idea that one language might be more 'expressive' than another?

There is of course the logical possibility that broad grammatical differences among languages might be derived from (or at least related to) differences in culture among the speakers of those languages. If such were the case, then any theory of language evolution would need to address the 'cultural need' for the development of whatever grammatical property is under discussion. The idea of an intimate grammar-culture link is mooted from time to time; indeed Everett has claimed that the lack of recursion in Pirahã is related to facts about

---

[2] Hence speakers are more likely to omit case markers when the information carried by those markers is predictable than when it is not (Kurumada & Jaeger 2015).

[3] For recent relevant discussion, see Piantadosi, Tily, & Gibson (2012), Kurumada & Jaeger (2015), and Fedzechkina, Jaeger, & Newport (2012).

Pirahã culture. However, since the pioneering work of Boas (1911/1963), linguists have overwhelmingly rejected the idea of such a link. To give a simple example, Russian and Chinese are both languages used as vehicles for modern science and technology, yet the former is quite complex morphosyntactically and the latter quite simple. On the other had, Bote (a language of Nepal) is of roughly the same complexity as Russian and Phuan (a language of Thailand) is of roughly the same complexity as Chinese, yet both are spoken in pre-technological cultures.

In sum, I have no problem with UG as a 'toolkit', but we need to acknowledge that speakers of languages having only hand tools manage to build the necessary structures as well as speakers that have power tools.

3.2 Is 'decremental' development attested? If so, how can that be reconciled with an incremental model?

Do some languages 'regress', in the sense that they lose features that are putatively evolutionarily 'advanced'? The answer is 'yes'. Starting with grammatical relations, it is apparently the case that some (if not all) languages that lack them now did have them at an earlier stage in their history. For example, the comparative evidence points to Acehnese being an innovation, as far as grammatical relations are concerned. Closely-related Austronesian languages do indeed have them (Ross 2002).

The same point can be made about the miniscule categorial inventory of Riau Indonesian. Related languages are much more 'developed' in terms of their categorial inventory and Gil (2005) has argued that the common ancestor of the Austronesian languages, spoken perhaps 5000 years ago, was substantially more complex than Riau Indonesian in many grammatical domains.

While we know nothing about the prehistory of Pirahã, the loss of recursive structures has been observed in a number of cases. Matsumoto (1988) calls attention to two ways of expressing in Japanese the proposition 'Although Taro is young, he does a good job'. One is by the simple conjunction of the two main propositions (2a), the other by use of the adversative subordinating suffix -*ga* (2b):

(2) a. Taro-wa wakai(-yo). Ga, yoku yar-u(-yo).
Taro-TOP young. but well do-PRES
'Taro is young. But he does a good job.'
b. Taro-wa wakai-ga, yoku yar-u(-yo)
Taro-TOP young, well do-PRES
'Although Taro is young, he does a good job.'

According to Matsumoto, paratactically-formed sentences such as (2a) have been recorded only since the seventeenth century, while the hypotaxis manifested in (2b) is observed much earlier. In other words, a recursive structure was lost.[4]

The loss of case markers is so well attested in Indo-European that no examples are necessary. The absence of case markers is often accompanied by rigid word order, but not always. On the one hand, a number of languages have no case marking, yet allow very flexible word order. Steele (1978) lists Classical Aztec, Karok, Achi, Wiyot, Tuscarora, Garadjari, and Maleceet-Passamoquoddy in this group and goes so far as to claim that 'the presence or absence of case marking has nothing to do with freedom of word order' (Steele 1978: 610).[5] Conversely, some languages, like Khamti, combine case marking with rigidity of word order (Mallinson & Blake 1981). That is even partly true for English, which case-marks pronouns, but does not thereby grant them freedom of occurrence (*I saw in the garden her).

So the question is how decremental development can be reconciled with an incremental model. Now, clearly, evolution is not directional and is certainly not unidirectional. Features that convey a fitness advantage at some point in evolutionary time might at a later point in time get lost: One thinks of flightless birds, believed to be descended from flying ancestors. But their loss of flight was compensated for by other factors, such as large size and speed of locomotion on the ground. Turning to the evolution of grammar, if the development of a certain feature represented an evolutionary advance, then why would a language lose that feature, in most cases without any obvious compensating gain elsewhere? In short, if UG is a toolkit, then why would speakers stop using a useful tool?

3.3 Are there alternative explanations for the development of morphosyntactic complexity that do not involve biological evolution?

The essential question here is whether it is *necessary* to posit biological evolution to account for the development of the grammatical features that Jackendoff considers to have arisen post-Protolanguage. The answer appears to be 'no'. The literature of diachronic linguistics is rich with examples of how these features have developed over time by means of 'natural' processes of language

---

[4]For more discussion of this and similar cases, see Traugott (1997).

[5]Peter Culicover (p. c.) suggests that the case and word order properties of these languages might follow from their polysynthetic morphology. This matter requires further investigation.

change.

Let us start with grammatical relations.[6] One common development is the transformation of topic markers into subjects. König (2008) gives a number examples of this change in various African languages. In the Australian language Bagandji, subjects developed from demonstratives (McGregor 2008) and in the Sahaptian languages from directional markers (Rude 1997). Objects as well have historical sources. For example, object markers can develop from 'take' verbs in constructions of the type 'take X and Verb (X)'. Such a development has been described for Mandarin Chinese (Li & Thompson 1981) and West African languages (Lord 1993).

Kulikov (2009) gives a detailed account of how case systems can arise over historical time. One common source of case morphemes is adpositions, where the development is attested in Indo-Aryan, Lithuanian, and Iranian. Other sources for case markers are adjectives and adverbials, and indexical elements such as pronouns and articles.

Even some of the features that Jackendoff considers to be evolutionarily ancient can arise through normal language change. The foremost of these is recursion. Deutscher (2000) documents the origins of finite recursive structures in Akkadian over the millennia in adverbial constructions and from the merging of two distinct arguments of a verb into one complement clause. The complementizers themselves arose from demonstratives. Another example of historically attested rise of recursivity is what happened in Sranan (Heine & Kuteva 2007). In its pidgin stage there were no formally marked relative clause structures at all. In a period of only a couple of hundred years, the demonstrative *disi* 'this' evolved into a relativizer with the concomitant development of full relative clause structures. If the Akkadian and Sranan cases are representative, it may not be necessary to appeal to biological evolution to explain the origins of recursivity, much less to assume that it was the major defining feature of the transition from non-language to language.

For both Jackendoff and Progovac, compounds are 'living fossils', harking back to the earliest stage of language, Jackendoff finding noun-noun compounds to be particularly ancient. But in fact compounds appear naturally over historical time, either as grammaticalizations of phrases[7] and in languages like Chinese that have developed increasingly reduced syllable structure and lack simplex accomplishment verbs (Huang 2015).

In sum, it is not easy to find examples of specific grammatical phenomena that *necessarily* arose via biological evolution. In all or almost all cases, well

---

[6]For an overview, see Cristofaro (2014).

[7]See Downing (1977) for numerous examples.

understood processes of diachronic change suffice to explain their appearance.

4 Conclusion

Anne Reboul has argued that enhanced communicative abilities could not have been the driving force in the origin and biological evolution of language. This paper has supported Reboul's hypothesis from the point of view of morphosyntax. It provides arguments to reject the incrementalist approach to language evolution proposed by Ray Jackendoff and others, in which the evolution of morphosyntax has proceeded in biologically-determined stages, each stage improving overall communicative efficiency.

References

Ackema P. & A. Neeleman (2002). "Effects of short term storage in processing rightward movement". In S. Noteboom, F. Weerman & F. Wijnen (eds). *Storage and computation in the language faculty.* Dordrecht: Kluwer. 219–256.
Aiello L.C. & R.I.M. Dunbar (1993). "Neocortex size, group size, and the evolution of language". *Current Anthropology* 34: 184–193.
Austin P. (2001). "Word order in a free word order language: The case of Jiwarli". In J. Simpson, D. Nash, M. Laughren & B. Alpher (eds). *Forty years on: Ken Hale and Australian languages.* Canberra: Pacific Linguistics. 305–323.
Berwick R. C. (1997). "Syntax facit saltum: Computation and the genotype and phenotype of language. *Journal of Neurolinguistics* 10: 231–249.
Bickerton D. (1981). *Roots of language.* Ann Arbor: Karoma.
Bickerton D. (1990). *Language and species.* Chicago: University of Chicago Press.
Bickerton D. (2009). *Adam's tongue: How humans made language, how language made humans.* New York: Hill and Wang.
Boas F. (1911/1963). *Introduction to the Handbook of American Indian languages.* Washington: Georgetown University Press.
Chomsky N. (1965). *Aspects of the theory of syntax.* Cambridge, MA: MIT Press.
— (1981). *Lectures on government and binding* (Vol. 9). Dordrecht: Foris.
— (1995). *The minimalist program.* Cambridge, MA: MIT Press.
— (2005). "Three factors in language design". *Linguistic Inquiry* 36: 1–22.

— (2007). "Approaching UG from below". In U. Sauerland & H.-M. Gärtner (eds). *Interfaces + recursion = language?*. Berlin: Mouton de Gruyter. 1–29.

— (2012). *The science of language: Interviews with James McGilvray*. Cambridge: Cambridge University Press.

Clark B. (2013). "Syntactic theory and the evolution of syntax". *Biolinguistics* 7: 169–197.

Cristofaro S. (2014). Competing motivation models and diachrony: What evidence for what motivations? In B. MacWhinney, A. Malchukov & E.A. Moravcsik (eds). *Competing motivations in grammar and usage*. Oxford: Oxford University Press. 282–298.

Culicover P.W. & R. Jackendoff (2005). *Simpler syntax*. Oxford: Oxford University Press.

DeGraff M. (ed). (1999). *Language creation and language change: Creolization, diachrony, and development*. Cambridge, MA: MIT Press.

Deutscher G. (2000). *Syntactic change in Akkadian: The evolution of sentential complementation*. Oxford: Oxford University Press.

Downing P. (1977). "On the creation and use of English compound nouns". *Language* 53: 810–842.

Dunbar R. (1996). *Grooming, gossip, and the evolution of language*. Cambridge, MA: Harvard University Press.

Durie M. (1985). *A grammar of Acehnese*. Dordrecht: Foris.

— (1987). "Grammatical relations in Acehnese". *Studies in Language* 11: 365–399.

Everett D.L. (2005). "Cultural constraints on grammar and cognition in Pirahã: Another look at the design features of human language". *Current Anthropology* 46: 621–646.

Falk D. (2004). "Prelinguistic evolution in early hominims: Whence motherese?" *Behavioral and Brain Sciences* 27: 491–541.

Fedzechkina M., T.F. Jaeger, & E.L. Newport (2012). "Language learners restructure their input to facilitate efficient communication". *Proceedings of the National Academy of Sciences* 109: 17897–17902.

Gil D. (2005). "Isolating-monocategorial-associational language". In H. Cohen & C. Lefebvre (eds). *Handbook of categorization in cognitive science*. Oxford: Elsevier. 347–379.

— (2007). "Creoles, complexity and associational semantics". In U. Ansaldo, S. Matthews & L. Lim (eds). *Deconstructing creole: New horizons in language creation*. Amsterdam: John Benjamins. 67–108.

— (2009). "How much grammar does it take to sail a boat?" In G. Sampson, D. Gil & P. Trudgill (eds). *Language complexity as an evolving variable* Oxford: Oxford University Press. 19–33.
Hale K. (1983). "Warlpiri and the grammar of nonconfigurational languages". *Natural Language and Linguistic Theory* 1: 5–47.
Heine B. & T. Kuteva (2007). *The genesis of grammar: A reconstruction.* Oxford: Oxford University Press.
Hewes G.W. (1973). "Primate communication and the gestural origin of language". *Current Anthropology, 14*: 5–25.
Huang C.-T.J. (2015). "On syntactic analyticity and parametric theory". In A. Li, A. Simpson & D. Tsai (eds). *Chinese syntax in a crosslinguistic perspective.* Oxford: Oxford University Press. 1–48.
Jackendoff R. (1983). *Semantics and cognition.* Cambridge, MA: MIT Press.
— (1990). *Semantic structures.* Cambridge, MA: MIT Press.
— (1992). *Languages of the mind: Essays on mental representation.* Cambridge, MA: MIT Press.
— (1997). *The architecture of the language faculty.* Cambridge, MA: MIT Press.
— (2002). *Foundations of language: Brain, meaning, grammar, evolution.* Oxford: Oxford University Press.
— (2010). "Your theory of language evolution depends on your theory of language". In R.K. Larson, V. Déprez & H. Yamakido (eds). *The evolution of human language: Biolinguistic perspectives.* Cambridge: Cambridge University Press. 19–33.
Kayne R.S. (1994). *The antisymmetry of syntax.* Cambridge, MA: MIT Press.
Kegl J., A. Senghas & M. Coppola (1999). "Creation through contact: Sign language emergence and sign language change in Nicaragua". In M. DeGraff (ed). *Language creation and language change: Creolization, diachrony, and development.* Cambridge, MA: MIT Press. 179–237.
Kibrik A.E. (1997). "Beyond subject and object: Toward a comprehensive relational typology". *Linguistic Typology* 1: 279–346.
König C. (2008). *Case in Africa.* Oxford: Oxford University Press.
Kulikov L. (2009). "Evolution of case systems". In A. Malchukov & A. Spencer (eds). *The Oxford handbook of case.* Oxford: Oxford University Press: 439–457.
Kurumada C. & T.F. Jaeger (2015). "Communicative efficiency in language production: Optional case-marking in Japanese". *Journal of Memory and Language* 83: 152–178.

Li C.N. & S.A. Thompson (1981). *Mandarin Chinese: A functional reference grammar*. Berkeley: University of California Press.

Lord C. (1993). *Historical change in serial verb constructions*. Amsterdam: John Benjamins.

MacDonald K. & W. Roebroeks (2013). "Neanderthal linguistic abilities: An alternative view". In R. Botha & M. Everaert (eds). *The evolutionary emergence of language: Evidence and inference*. Oxford: Oxford University Press. 97–117.

Mallison G. & B. Blake (1981). *Language Typology: Cross-linguistic Studies in Syntax*. Amsterdam: North-Holland.

Matsumoto Y. (1988). "From bound grammatical markers to free discourse markers: History of some Japanese connectives". In S. Axmaker, A. Jaisser, & H. Singmaster (eds). *Proceedings of the 14th Annual Meeting of the Berkeley Linguistics Society*. 340–351.

McGregor W.B. (2008). "Indexicals as sources of case markers in Australian languages". In F. Josephson & I. Söhrman (eds), *Interdependence of diachronic and synchronic analyses*. Amsterdam: John Benjamins. 299–321.

Miller G. (2001). *The mating mind: How sexual choice shaped the evolution of human nature*. New York: Doubleday.

Millikan R. G. (2005). *Language: A biological model*. Oxford: Clarendon Press.

Mobbs I. (2014). *Minimalism and the design of the language faculty*. Unpublished Ph. D. thesis, University of Cambridge.

Newmeyer F.J. (1991). "Functional explanation in linguistics and the origins of language". *Language and Communication* 11: 3–28.

Nordlinger R. (1998). *Constructive case: Evidence from Australian languages*. Standord, CA: CSLI.

Piantadosi S.T., H. Tily, & E. Gibson (2012). "Word lengths are optimized for efficient communication". *Proceedings of the National Academy of Sciences* 108: 3526.

Pinker S. (1989). *Learnability and cognition: The acquisition of argument structure*. Cambridge, MA: MIT Press.

Pinker S. & P. Bloom (1990). "Natural language and natural selection". *Behavioral and Brain Sciences* 13: 707–784.

Power C. (1998). "Old wives' tales: The gossip hypothesis and the reliability of cheap signals". In J.R. Hurford, M. Studdert-Kennedy & C. Knight (eds), *Approaches to the evolution of language: Social and cognitive bases*. Cambridge: Cambridge University Press. 111–129.

Primus B. (1999). *Cases and thematic roles: Ergative, accusative, and active.* Tübingen: Max Niemeyer.

Progovac L. (2015). *Evolutionary syntax.* Oxford: Oxford Uiversity Press.

Reboul A.C. (2015). "Why language really is not a communication system: A cognitive view of language evolution". *Frontiers in Psychology* 6: 1–12.

Ross M. (2002). "The history and transitivity of western Austronesian voice and voice-marking". In F. Wouk & M. Ross (eds). *The history and typology of western Austronesian voice systems.* Canberra: Pacific Linguistics. 17–62.

Rude N. (1997). "On the history of nominal case in Sahaptian". *International Journal of American Linguistics* 57: 24–50.

Sandler W., I. Meir, C. Padden, & M. Aronoff (2005). "The emergence of grammar: Systematic structure in a new language". *Proceedings of the National Academy of Sciences* 102: 2661–2665.

Scott-Phillips T. (2015). *Speaking our minds: Why human communication is different, and how language evolved to make it special.* Basingstoke: Plagrave Macmillan.

Steele S. (1978). "Word order variation: A typological study". In J.H. Greenberg, C.A. Ferguson & E.A. Moravcsik (eds). *Universals of human language* (Vol. 4: Syntax). Stanford, CA: Stanford University Press. 585–623.

Talmy L. (1983). "How language structures space". In H.L. Pick & L.P. Acredolo (eds). *Spatial orientation: Theory, research, and application.* New York: Plenum Press. 585–623

— (1988). "Force dynamics in language and cognition". *Cognitive Science* 12: 49–100.

Tomasello M., M. Carpenter, J. Call, T. Behne & H. Moll (2005). "Understanding and sharing intentions: The origins of cultural cognition". *Behavioral and Brain Sciences* 28: 675–735.

Traugott E.C. (1997). "The role of the development of discourse markers in a theory of grammaticalization". In L. v. Bergen & R.M. Hogg (eds). *Papers from the 12th International Conference on Historical Linguistics.* Amsterdam: John Benjamins.

Van Valin R.D. & R.J. LaPolla (1997). *Syntax: Structure, meaning, and function.* Cambridge: Cambridge University Press.

Washburn S.L. & C.S. Lancaster (1968). "The evolution of hunting". In R.B. Lee & I. DeVore (eds). *Man the hunter.* Chicago: Aldine. 293–303.

Wilkins W.K. & J. Wakefield (1995). "Brain evolution and neurolinguistic preconditions". *Behavioral and Brain Sciences* 18: 161–226.

Part III

Cognitive Science and Philosophy of Mind

Chapter 11

Stop Moving and You Will Understand. Selective Weakening of the N400 to Hand-Action Words during Hand Movements

Tatjana A. Nazir, Raphaël Fargier, Evgueni Douissembekov & Yves Paulignan

Abstract Previous research has shown that processing language that describes motor actions activates areas of our brain that plan and program the execution of motor actions. Using EEG and behavioral measures, we show that executing an arm/hand movement selectively affects the processing of verbal action description. This finding implies that language and motor functions share common resources and suggest that the processing of action words profits from the participation of motor areas of the brain.

Keywords Embodied language · Action words · Motor action · EEG

When listening to verbal descriptions of body movements or actions (e.g. a soccer game on the radio), areas of our brain that are responsible of planning and executing motor actions become active even though we do not perform the movements ourselves (see Hauk et al. 2004; Simmons et al. 2007 for reviews). This and similar observations in the domain of perception (e.g. color words activate brain areas responsible for the perception of colors) have triggered heated arguments as to whether human language comprehension could involve brain systems used for perception and action (Mahon & Caramazza 2008; Martin 2007; Pulvermüller 1999). Using a simple test, we show here that correct understanding of verbal action descriptions might indeed require the participation of motor-areas of the brain.

If the hypothesis that language-induced activity in typical motor brain areas contributes to the understanding of verbal actions descriptions is correct, the ability to seize the meaning of these descriptions should decline when motor areas are occupied with planning and executing movements. By contrast, understanding of verbal descriptions that do not describe actions should not be affected by the movement. As we will show here, the analysis of semantic judgment errors during sentence comprehension and the analysis of an electrophysiological marker that indexes the efforts of integrating the meaning of a word in its sentence context (the N400 component) confirm this assumption.

Evoked response potentials (ERPs) were recorded from 27 native speakers of French while they were listening to sentences that ended with a concrete noun or an action verb. This last word was either semantically congruent or anomalous within the sentence context (details in the **Methods** section below). In one condition participants simply listened to the sentences and replied with "yes" or "no" depending on whether they considered that a sentence was semantically congruent. In a second condition, along with the language-task, they were required to continuously grasp and place a cylindrical object in one of two holes. A random illumination of the holes indicated the target of the movement. Semantic judgments and the N400 component (a negative deflection in the ERP over central-parietal sites approximately 400 ms after word onset) were recorded. The N400 amplitude is typically larger for sentences with semantically anomalous compared to semantically congruent contents. This difference is referred to as the "N400 effect" (Kutas & Federmeier 2000). If understanding action words requires the participation of motor brain areas, the concurrent movement-task should lead to semantic judgment errors for sentences that end with verbs but not for those that end with nouns. Moreover, the magnitude of the "N400 effect" in the verb-condition should diminish because if the sentence final word is not processed, congruous and anomalous sentences become identical. For the noun-condition, no such movement-induced attenuation should be observed.

Semantic judgment errors for the verb-condition (but not for the noun-condition) were slightly but significantly higher when participants performed the movement-task (**Fig. 11.1a**) (Movement × Word class $F(1,26) = 10.33$, $p < .01$). ERPs were also differentially affected by the movement-task depending on the sentences final word. For the noun-condition the movement had no impact on the magnitude of the "N400 effect" (**Fig. 11.1b-d**). By contrast, for the verb-condition, the "N400 effect" diminished when a movement was required (**Fig. 11.1e-g;** statistical details in **Table 11.1** below). To assure that the selective impairment in the verb-condition was specific to the movement and not simply due to the double-task, we asked another 27 participants to perform the same language-task but replaced the movement-task with counting (counting how often the holes in the experimental device were illuminated twice in a row). The attention demanding counting-task should affect the understanding of all sentences, whether they end with verbs or nouns. Accordingly, in contrast to the movement-task, which preserved sentences that ended with nouns, the counting-task affected the "N400 effect" in the verb- and in the noun-condition (**Fig. 11.2a-g** and **Table 11.2** below).

Moving while listening to language can thus selectively impair our understanding of verbal descriptions of actions without affecting our understanding of verbal descriptions of non-actions. This transient attenuation is expected if two tasks compete for "common brain resources" (Boulenger et al. 2006). When motor brain areas are occupied with planning and executing actions, less resource is left for language processes. Related phenomena are also seen in pathological conditions. Patients that suffer from Parkinson's disease (a neurological disorder that affects motor functions) perceive action verbs less well than concrete nouns. This selective deficit disappears, however, with treatment that restores motor functions (Boulenger et al. 2008). As argued by others (Pulvermüller 1999; Keysers & Gazzola 2006) and demonstrated experimentally (e.g Lahav et al. 2007; Goldfield 2000; Cooks et al. 2008) the sharing of cortical resources between functions probably emerges through association learning. In (Lahav et al. 2007) non-musicians were trained to play a piece of music on the piano. Once acquired, mere listening to the learned piece activated motor brain areas. In the case of language, the co-occurrence of actions and their verbal descriptions during language acquisition (verbs in maternal speech often occur to elicit children's actions, see Goldfield 2000) may be the basis for the here-observed association. While it is unlikely that all we know about action words is stored in motor areas, integrating these areas into the

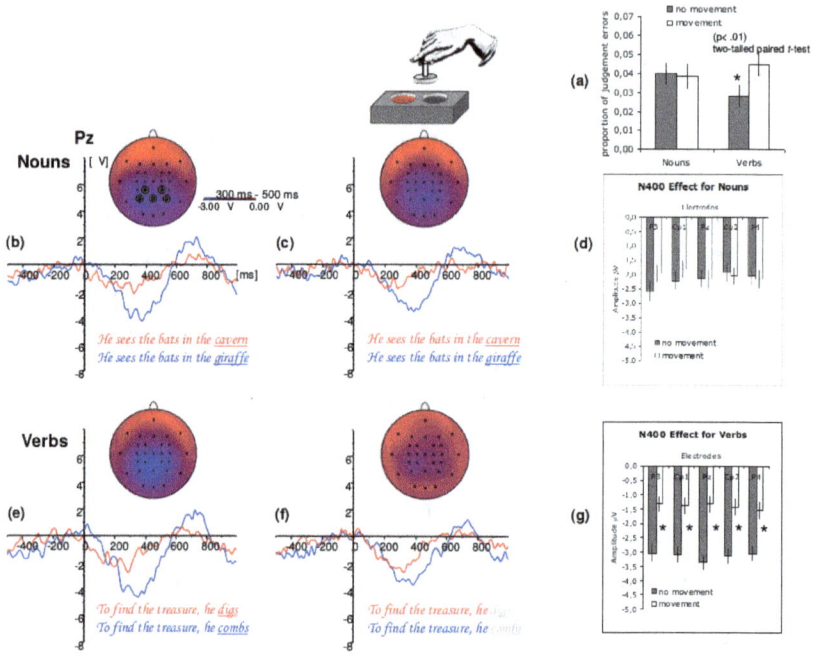

Figure 11.1: **(a)** Proportion of semantic judgment errors for sentences that ended with nouns and verbs. **(b-c; e-f)** Grand average ERPs at a central-parietal electrode (Pz), time-locked to the onset of the final noun (top) and verb (bottom). Congruent sentences in red, ambiguous sentences in blue. (b and e) Conditions without and (c and f) with movement. Spatial cartography shows the difference between the N400 amplitude in the two conditions averaged over a 300-500 ms latency window following word onset. White circles in (b) indicate electrodes of interest (left top to bottom right: Cp1, Cp2, P3, Pz, P4). **(d and g)** Magnitude of the "N400 effect" (difference between the largest negative peak within the 300-500ms latency window in the two conditions) at the five electrodes. Data are plotted with standard errors. Two-tailed paired t-tests for each electrode revealed no impact of the movement for sentences that ended with nouns. For sentences that ended with verbs the movement-task diminished the "N400 effect". (*) indicate p-values < .05.

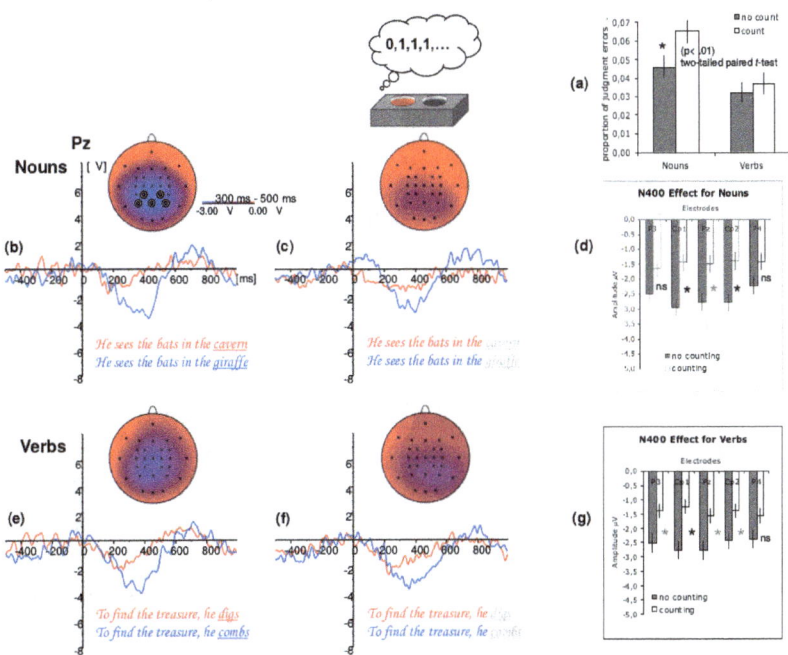

Figure 11.2: (a) Proportion of semantic judgment errors for sentences that ended with nouns or verbs. (b-c; e-f) Grand average ERPs at a central-parietal electrode (Pz), time-locked to the onset of the final noun (top) and verb (bottom). Congruent sentences in red, ambiguous sentences in blue. (b and e) Conditions without, (c and f) with counting. (d and g) Magnitude of the "N400 effect" at the five electrodes of interest. The counting-task diminished the "N400 effect" in the same way for sentences that ended with verbs as well as those that ended with nouns. The effect of counting on the "N400 effect" was not very stable though because paired t-tests revealed p-values between 0.05 (black (*)) and 0.10 (grey (*)), and no significant (ns) differences at some electrodes.

brain network that process language may expand verb understanding to cover knowledge of how an action is executed. This "embodied" way of representing meaning could serve the better retention of words and possibly help consolidating newly acquired concepts as even simple gesturing seems to make learning last (Goldin-Meadow 2009).

Method

Participants   Twenty-seven healthy right-handed French native volunteers participated in each of the two experiments (age range 20-33 years). Participants reported no pre-existing physical or neurological conditions. All participants provided written informed consent and were free to withdraw from the experiment at any time.

Materials   A total of 35 nouns and 35 verbs, controlled for frequency, number of letters, number of syllables, bi- and trigram frequency in the infinitive form ("Lexique"; see New et al. 2001), and for which a pilot lexical decision experiment with 10 participants showed equivalent response latencies, served as target words at the end of sentences. All verbs denoted actions performed with the hand/arm (e.g. write, throw), while nouns referred to imaginable concrete entities that cannot be manipulated, (e.g. iceberg, canyon). Words that could be used as nouns and verbs were excluded from the selection.

For each word class, a set of two identical 35 spoken sentences that ended with a target word was created, leading to a total of 70 sentences per word class. The attribution of the final critical word to a sentence context was such that one of the two identical sentences was semantically congruent with the final word, and one was not. Each target word occurred once within a congruent context and once within an anomalous context. Sentences were recorded with a normal speaking rate (female native speaker) on a digital recorder. The overall duration of the sentences and vocal parameters (e.g. pitch, loudness) were controlled using Cubase SX computerised program.

Procedure   A close-fitting cap containing 32 electrodes was placed on participants scalp, and a small quantity of water-soluble conductive gel was used to ensure satisfactory impedance. Participants were comfortably seated in a dimly lit room, 60 cm from the screen. Participants were instructed to relax and to minimise eye movement, muscle tension, and gross body movement during the task.

Participants completed two conditions. Each trial consisted of sentence presentation, followed by a response interval. In one condition participants simply listened to the 140 sentences that were presented through speakers. Following each sentence participants were required to answer with "yes" if they considered that the sentence was semantically congruent and with "no" if they considered that the sentence was anomalous. The verbal response was recorded by the experimentator who triggered the next trial. If participants did not provide an answer within 6s after the end of the sentence, the next trial was generated automatically and the response was counted as omission. In the second condition of the "movement" experiment, participants were presented with the same spoken sentences. However, concurrently to the language-task, participants were asked to perform a motor-task that required grasping and placing a small cylindrical object with a flat translucent base in one of two flat holes (depth: 10 mm, diameter: 40 mm). The holes were spaced 5 cm apart (center to center) and were briefly and randomly illuminated from underneath to indicate the target of the movement. The motor-task, which required participants to produce a continual movement with the forearm while listening to the sentences (54 grasps/min) was performed non-stop throughout the entire block. The order of the two conditions was counterbalanced across participants and took approximately 25 minutes to complete. In the "counting" experiment, participants were requested to silently count (in a binary zero-or-one mode) the number of times the holes were illuminated twice in a row to determine whether this number was odd (one) or even (zero).

EEG data acquisition and analysis Electroencephalographic (EEG) data were recorded using BrainAmp amplifiers (Brain Vision recorder software, Brain Products GmbH, Munich, Germany). EEG was recorded from 32 scalp sites using the international 10-20 system (American Encephalographic Society, 1994), with a forehead ground and impedance at 10 K or less at the start of the recording session. All scalp sites were referenced to the left mastoid (M1). Eye movements were monitored using horizontal electro-oculogram (EOG) with a bipolar recording from electrodes placed at the left and right outer canthi of the eyes. Eye blinks and vertical eye movements were recorded using an electrode placed below the left eye. Standard Ag/AgCl electrodes were used.

EEG activity was analysed using BrainAmp Analyze software. The EEG was first filtered (band-pass 1-30 Hz, 48dB/octave) and data were EOG corrected to account for ocular artifact (Gratton et al., 1983). Data were grouped according to the four sentence conditions (Noun-anomalous, Noun-coherent, Verb-anomalous, and Verb-coherent) and epoched from 500 ms before to 1000 ms after acoustic onset of the critical word. Epochs in which the EEG or EOG exceeded ±200 µV were rejected. Epochs were then baseline corrected (200 ms pre-stimulus baseline) and averaged for each conditions. This analysis was performed for the movement (counting) and no movement (no counting) trials separately. Therefore, there were 8 conditions in total per experiment. The Pz electrode as well as four electrodes surrounding this region (Cp1, CP2, P3, P4) was defined as the region of interest. The N400 component was defined as the largest negative peak occurring within the latency window of 300-500 ms post-stimulus onset. Peak amplitude was measured relative to the pre-stimulus baseline.

Results

Movement experiment

ERP data A three-way ANOVA on the average maximum peak amplitudes (µV) at the Pz electrode, which crossed Word Class (Nouns, Verbs), Coherence (Coherent, Anomalous) and Movement (no movement, movement) revealed a significant effect of the main factor Coherence ($F(1,26) = 55.06$, $p < .0001$), a significant interaction between Coherence and Movement ($F(1,26) = 5.74, p = .024$) and a significant triple interaction between the three main factors ($F(1,26) = 5.41, p = .028$). Separate analyses for nouns revealed significant effects of the factors Coherence and Movement with no interaction between them (**Table 11.1**). That is, the concurrent movement reduced the overall amplitude of the N400 component relative to the no-movement condition but the magnitude of the "N400 effect" (i.e., the difference of the amplitude of the N400 component in the coherent and anomalous sentence contexts) was not affected by the movement. For verbs, there was a significant effect of Coherence, no effect of Movement but a significant interaction between Coherence and Movement. In contrast to nouns, in the verb-condition the concurrent movement reduced the amplitude of the N400 component in the anomalous sentence context but increased it in the coherent sentence context. As a consequence, the magnitude of the "N400 effect" diminished. The same analysis was performed for the four electrodes in the vicinity of the Pz electrode. For all electrodes a significant effect of the main factor Coherence (all $p < .01$) and a significant interaction between Coherence and Movement were found (all $p < .05$). Although the triple interaction between the three main factors was only significant or marginally significant for the two electrodes over the right hemisphere ($F(1,26) = 5.21, p = .031$ for CP2; $F(1,26) = 3.02, p = .094$

for P4), separate analyses for nouns and verbs revealed essentially the same pattern of results at all electrodes (**Table 11.1**).

Counting experiment

Behavioral data Semantic judgment errors for the verb- and noun-conditions were slightly higher when participants performed the counting-task (**Fig. 11.1a**; effect of the main factor Counting, $F(1,26) = 6.628, p = .016$). A significant interaction between the factors Counting × Word class ($F(1,26) = 4.459, p = .044$) was also found, indicating that counting had a stronger effect on the noun-condition. Separate analyses for nouns and verbs revealed a significant effect of counting for the noun-condition only ($p < .01$).

ERP data The counting-task diminished the "N400 effect" in the verb- as well as in the noun condition (**Fig. 11.2b-g**). The three-way ANOVA revealed a significant effect of the main factor Coherence for all electrodes (all $p < .01$) and a significant interaction between Coherence and Counting for all but electrode P4 (all $p < .05$). The triple interaction (Word Class, Coherence and Counting) was not significant at any electrode. Separate analyses for nouns and verbs (**Table 11.2**) showed similar effects of the counting-task on the two classes of words.

		Electrodes									
		P3		Cp1		Pz		Cp2		P4	
		F	p	F	p	F	p	F	p	F	p
Nouns	Movement	2,73	ns	7,26	,012*	5,96	,022*	10,19	,004*	6,21	,019*
	Coherence	39,32	,0001*	29,62	,0001*	25,37	,0001*	26,41	,0001*	26,82	,0001*
	M×C	1,72	ns	0,6	ns	0	ns	0,04	ns	0,04	ns

		Electrodes									
		P3		Cp1		Pz		Cp2		P4	
		F	p	F	p	F	p	F	p	F	p
Verbs	Movement	0,04	ns	0,9	ns	0,3	ns	1,31	ns	0,29	ns
	Coherence	62,66	,0001*	79,97	,0001*	79,01	,0001*	70,5	,0001*	60,59	,0001*
	M×C	5,84	,023*	6,9	,014*	9,34	,005*	8,3	,008*	6,68	,015*

Table 11.1: *F*- and *p*-values for the effects of movement and sentence coherence observed at the 5 tested electrodes for sentence that ended with nouns or verbs

		Electrodes									
		P3		Cp1		Pz		Cp2		P4	
		F	p	F	p	F	p	F	p	F	p
Nouns	Counting	,054	ns	,05	ns	0,57	ns	,034	ns	,237	ns
	Coherence	53,74	**,0001***	45,34	**,0001***	48,9	**,0001***	37,39	**,0001***	33,35	**,0001***
	C×C	1,92	ns	5,59	**,026***	3,39	,077	4,30	**,048***	1,44	ns

		Electrodes									
		P3		Cp1		Pz		Cp2		P4	
		F	p	F	p	F	p	F	p	F	p
Verbs	Counting	0,18	ns	,20	ns	0,013	ns	,66	ns	,114	ns
	Coherence	23,86	**,0001***	34,04	**,0001***	41,81	**,0001***	41,25	**,0001***	36,4	**,0001***
	C×C	3,34	,079	5,98	**0,22***	3,18	,086	2,84	,10	1,45	ns

Table 11.2: *F*- and *p*-values for the effects of counting and sentence coherence observed at the 5 tested electrodes for sentence that ended with nouns or verbs

References

American Electroencephalographic Society (1994). "American Electroencephalographic Society Guidelines for Standard Electrode Position Nomenclature". *Journal of Clinical Neurophysiology* 8(2): 200–202.

Boulenger V., A.C. Roy, Y. Paulignan, V. Deprez, M. Jeannerod & T.A. Nazir (2006). "Cross-talk betzeen Language Processes and Overt Motor Behavior in the First 200 msec of Processing". *Journal of Cognitive Neuroscience* 18(10): 1607–1615.

Boulenger V., L. Mechtouff, S. Thobois, E. Broussolle, M. Jeannerod, T.A. Nazir (2008). "Word Processing in Parkinson's disease is impaired for action verbs but not for concrete nouns". *Neuropsychologia* 46: 743–756.

Cook S.W., Mitchell Z., & Goldin-Meadow S. (2008). "Gesturing makes learning last". *Cognition* 106(2): 1047–1058.

Goldfield B.A. (2000). "Nouns before verbs in comprehension vs. production: the view from pragmatics". *Journal of Child Language* 27: 501–520.

Goldin-Meadow S. (2009). "How gesture promotes learning throughout childhood". *Child Development Perspectives* 3(2): 106–111.

Gratton G., Coles M.G. & Donchin E. (1983). "A new method for off-line removal of ocular artifact". *Electroencephalography and Clinical Neurophysiology* 55(4): 468–484.

Hauk O., Johnsrude I. & Pulvermüller F. (2004). "Somatotopic representation of action words in human motor premotor cortex". *Neuron* 41(2): 301–307.

Keysers C. & Gazzola V. (2006). "Towards unifying neural theory of social cognition". *Progress in Brain Research* 156: 379–401.

Kutas M. & Federmeier K.D. (2000). "Electrophysiology reveals semantic memory use in language comprehension". *Trends in Cognitive Sciences* 4(12): 463–470.

Lahav A. Saltzman, E. & Schlaug, G. (2007). "Action Representation of Sound: Audiomotor Recognition Network While Listening to Newly Acquired Actions". *Journal of Neuroscience* 27(2): 308–314.

Mahon B.Z. & Caramazza A. (2008). "A critical look at the embodied cognition hypothesis and a new proposal for grounding conceptual contentt't'. *Journal of Physiology Paris* 102(1–3): 59–68.

Martin A. (2007). "The Representation of Object Concepts in the Brain". *Annual Review Psychology* 58: 25–45.

New B. et al. (2001). "A lexical database for contemporary french: LEXIQUE™". *L'Année psychologique* 101(3–4): 447–462.

Pulvermüller F. (1999). "Words in the brain's language". *Behavioral and Brain Sciences* 22(2): 253–279.

Simmons W.K. et al. (2007). "A common neural substrate for perceiving and knowing about color". *Neuropsychologia* 45(12): 2802–2810.

Chapter 12
Should Embodied Cognitive Science Go Radical? A Hint from Music

Pierre Saint-Germier

Abstract According to the Radical Embodied Cognitive Science portrayed by Anthony Chemero, (i) representational and computational views of embodied cognition are mistaken (ii) embodied cognition is to be explained with tools including dynamical systems theory and (iii) which do not posit mental representations. The goal of this paper is to critically examine this proposal. The nuanced proposal of Chemero concerning mental representations is examined. A diagnosis proposed by Reboul, according to which the debate between eliminativist and realists about mental representation in the philosophy of Embodied Congnitive Science threatens to be merely verbal is explored and, to a certain extent, vindicated in the case of Chemero. It is also argued, on the basis of an example drawn from rhythm cognition, that REC has trouble accounting for some dynamical explanations which indispensably involve representations.

Keywords Embodied cognition · Mental representation · Dynamical Systems Theory · Rhythm perception · Resonance Theory

1 Introduction

The insight that the capacities and achievements of cognitive agents is intimately dependent upon features of their bodies is central to the research program known as "Embodied Cognition".[1] This insight has lead cognitive scientists working in this area to introduce new approaches and methods in the explanatory toolbox of cognitive science.

The unity of a label should not conceal a diversity of views concerning the sense in which cognition is embodied, and the best way to study cognition as an embodied capacity. Shapiro usefully distinguishes three distinct projects that have been conducted under the common flag of "Embodied Cognition". One effort has been to defend a *Conceptualization* hypothesis, according to which "an organism's understanding of the world—the concepts it uses to partition the world into understandable chunks—is determined in some sense by the properties of its body and sensory organs" (Shapiro 2011: 68).[2] Another one has been to promote the *Replacement* of the computational and representational tools of classical cognitive science in favour of non-computational and non-representational explanatory frameworks, such as dynamical systems theory.[3] Yet another project is to argue for a *Constitution* hypothesis, according to which "mental activity includes the brain, the body, and the world, or interactions among these things" (Shapiro 2011: 68).[4]

In this chapter, I focus on the second trend, as recently explored by Anthony Chemero in his book *Radical Embodied Cognition* (2009). Chemero's proposal of a Radical Embodied Cognition program clearly calls for a *Replacement* of the classical representational and computational cognitive science. A distinctive aspect of REC, compared to other approaches of Embodied Cognition, is its eliminative stance towards mental representations. Not all followers of the Embodied Cognition program are willing to throw away mental representations: the Conceptualization hypothesis, as defined by Shapiro, is intrinsically representationalist and the Constitution hypothesis can accommodate mental representations, provided they extend somehow in the body and perhaps also in the environment (e.g. Clark 1997, 2008). I propose here to evaluate Chemero's case for the elimination of mental representation from the

---

[1] See Shapiro (2011) for a lucid introduction to this field and Shapiro (2014) for a recent state of the art.

[2] The influential book of Varela, Rosch and Thomson (1991) falls within the approach, as well as Lakoff and Johnson's work (1980, 1999).

[3] Esther Thelen and Linda Smith have been influential advocates of this approach in developmental psychology (Thelen & Smith 1994).

[4] Andy Clark has nicely synthetized this view in his writings (e.g. Clark 1997, 2008).

program of Embodied Cognition, in connection with some critical comments offered by Reboul (2011). I will examine, in particular, a recent dynamical explanation of rhythm perception because it is *prima facie* a good example to motivate the kind of radical program promoted by Chemero. However, the same example raises important difficulties for REC's distinctly radical claims, or so I shall argue.

2 Chemero on Radical Embodied Cognitive Science

Let us start by reviewing the main tenets of Chemero's radical program for Embodied Cognitive Science, with a special attention to the theme of representation.

REC amounts to two positive claims and one negative claim:

> **Radical embodied cognition, claim 1** Representational and computational views of embodied cognition are mistaken.
>
> **Radical embodied cognition, claim 2** Embodied Cognition is to be explained via a particular set of tools T, which includes dynamical systems theory.
>
> **Radical embodied cognition, claim 3** The explanatory tools in set T do not posit mental representations.
>
> ...I hereby define radical embodied cognitive science as the scientific study of perception, cognition, and action as necessarily embodied phenomenon, using explanatory tools that do not posit mental representations. It is cognitive science without mental gymnastics. (Chemero 2009: 29)

Chemero sees REC as a descendant of the anti-representationalist traditions of American naturalism (Dewey) and of Gibsonian ecological psychology, rather than a recent bifurcation within the representational and computational mainstream of cognitive science, consisting in emphasizing how information processing can be offloaded from the brain into the body and even into the environment itself. The Radicalism of REC lies primarily in its clear rejection of representation and computation from the explanation of cognitive phenomena, which distinguishes it from more moderate approaches to Embodied Cognition (e.g. Clark 1997, 2008).

Interestingly, Chemero does not *argue* for Claim 1, mostly because he doubts that philosophical arguments against general explanatory strategies in cognitive science are dialectically effective. According to him, they indicate a

220 SHOULD EMBODIED COGNITIVE SCIENCE GO RADICAL?

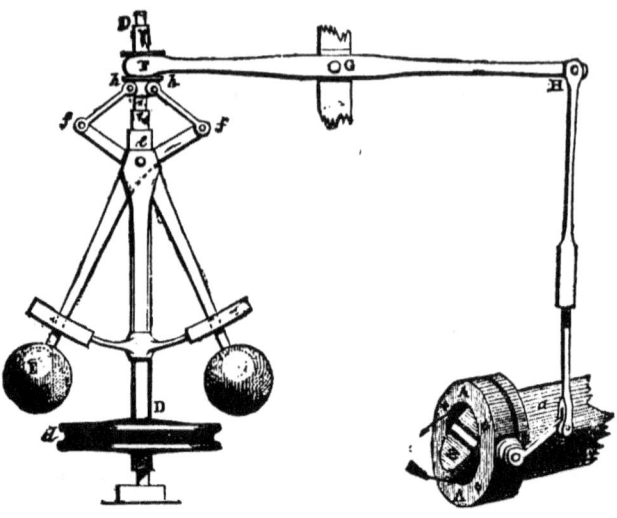

Figure 12.1: Watt's centrifugal governor. From Routledge (1900: 6).

state of immaturity of the field rather than make steps towards progress. Yet it is a basic tenet of REC that representation and computation are at least unnecessary, if not detrimental, to an adequate understanding of cognitive phenomena. REC promotes instead non-representational and non-computational forms of explanation.

In virtue of its second claim, REC privileges a specific way to think about cognitive systems and explain their behaviour, namely as dynamical systems rather than as digital computers (Pylyshin 1984). Instead of analyzing cognitive systems as sets of discrete representations modified step by step according to algorithmic rules—what Chemero sarcastically calls "mental gymnastics"—REC promotes a view of cognitive systems as systems governed by nonlinear differential equations linking continuously features of the brains of cognitive agents to features of their body and immediate environment. Rather than Turing machines, REC prefers Watt's centrifugal governor as the toy example of a cognitive agent (see Figure 12.1).[5]

Let us consider this example in some detail, to get a better insight into the dynamical way of thinking about cognition. In order to adapt the amount of fuel so that the speed of an engine remains constant, Watt's governor does not need to *compute* the appropriate amount of fuel on the basis of a *representation*

---

[5] See also van Gelder (1995) for a presentation of this example and an elaboration of the point that the dynamical explanation of the governor involves neither representation nor computation.

of the current speed of the engine, but connects dynamically the two quantities themselves via a physical mechanism satisfying the following differential equations:

$$\frac{d^2\theta}{d^2t} = (n\omega)^2 \cos\theta \sin\theta - \frac{g}{l}\sin\theta - r\frac{d\theta}{dt} \qquad (12.1)$$

where θ is the angle of the arms, n is a gearing constant, ω is the rotational speed of the engine, g is a constant for gravity, l is the length of the arms, and r is a constant of friction at hinges, and

$$I\frac{d\omega}{dt} = \alpha\cos\theta - \alpha\cos\theta_0 + F_1 - P \qquad (12.2)$$

where I is the moment of inertia of the flywheel, α is a positive constant of proportionality, P is the torque due to the load on the flywheel, ω_0 is the desired constant speed, and F_1 is the torque due to the action of the steam.[6]

This toy example allows to introduce at this point two important concepts. The first one is that of *coupling*. At the most general level, coupling occurs between two objects whenever they interact in such a way that their behaviour is best understood as part of a single system when the behaviour of each is a function of the behaviour of the other. In the Watt governor, the angle θ of the arms can be seen as coupled with the speed of the engine. One can see the angle of the arm a part of the system responsible for the speed of the engine and one can see the speed of the engine as a part of the system responsible for the opening of the valve; or one can see both as states within a single dynamical system.

The second concept is that of nonlinearity. The equations 12.1 and 12.2 have the important mathematical property to be *nonlinear*: for each equation, the quantity on the left side is not a *linear* function of the constant and variable quantities on the right side. Nonlinearity matters for a number of reasons.[7] One of them is that the effect of a sum of stimuli cannot be analyzed as the sum the the effects of each stimulus. This means that the global behaviour of the system cannot be analyzed as a sum of local behaviours. For this reason, Chemero takes the existence of a nonlinear coupling between the brain, the body and the environment of a cognitive agent to support the view that cognition extends to the body and even into the environment itself.[8]

---

[6] For more details about the derivation of the equations, see, e.g., Beltrami (1987: 162–164).
[7] See, e.g., the first chapter of (Strogalz 1994) for a sense of these reasons.
[8] Note that even if many interesting dynamical systems—especially those that include cognitive agents—instantiate both coupled states and nonlinear dynamics, the two concepts are in

Beer's work in artificial intelligence (2003) serves as a useful example of the way dynamical systems can account for capacities which, unlike the speed-controlling capacity of Watt's governor are uncontroversially *cognitive*. Beer studies an artificial agent which, via a genetic algorithm, evolved to discriminate between round and rectangular objects falling vertically above it, and to adapt its behaviour accordingly, by grasping the round objects and avoiding the rectangular objects. The evolved agent is therefore capable of categorical perception. Beer models the whole system comprising the artificial neurons of the agent, his body and the falling objects as a set of nonlinear differential equations and relies on its dynamics to explain how this simplified artificial agent achieves categorical perception. What the dynamical approach reveals here is the importance of the relations at each time between the neurons of the agent, the position of its body and the the position of the object to be discriminated.

According to claim 3, it is not necessary to posit representational states to explain the behaviour of cognitive agents when they are treated as parts of dynamical systems. For example, in the case of Beer's evolved artificial agents, assigning a representational content to each neurons, or collectively to the whole network does not take us any closer to a good explanation of the behaviour of the agent. The coupling between the states of the neurons, the states of the agent's body and the state of the object to be discriminated is already sufficient to provide the explanation.

Chemero's take on mental representation is however more nuanced than what the condensed formulation of claim 3 suggests.

3 Chemero on representation: yes and no

Chemero in fact distinguishes between a *Metaphysical Claim*, according to which there *are* no representations in cognitive systems and an *Epistemological Claim* according to which representations are not *involved in the best explanation* of the behaviour of cognitive systems. Let us see more precisely how each of these two claims are justified.

Chemero's preferred definition of representation is based on Millikan's teleofunctional theory of representation (1984, 1993). Here is Chemero's official definition:

---

principle separable. For example, the behaviour of coupled harmonic oscillators can be described by a linear dynamical system (e.g Pain 2005: 79–81) and the behaviour of a single, non coupled, simple pendulum is modelled by a nonlinear dynamical equation (e.g. Strogatz 1994: 6–7)).

A feature R_0 of a system S is a Representation for S if and only if:

(R1) R_0 stands between a representation producer P and a representation consumer C that have been standardized to fit one another.

(R2) R_0 has as its function to adapt the representation consumer C to some aspect A_0 of the environment, in particular by leading S to behave appropriately with respect to A_0, even when A_0 is not the case.

(R3) There are (in addition to R_0) transformations of R_0, R_1,..., R_n, that have as their function to adapt the representation consumer C to corresponding transformations of $A_0, A_1,...,$ A_n. (Chemero 2009: 50–51)

This definition is teleofunctional in that it assigns to representations a *function* to adapt parts of a cognitive system to some aspects of its environment. This teleological aspect enables to account naturalistically for misrepresentation: a false representation is just a representation which does not fulfill its function (R_2). This requires to distinguish within a cognitive agent the parts which produce representations from those which consume them (R_1). In addition, each representation perform its function as part of a *system* of representaion, such that there is a mapping between variations in the representations and variations in the features of the environment represented. For example, in a system of representations containing negation, the negation of a representation R represents the negation of what R represents (R_3).

This definition of representation does not rule out the possibility that a dynamical system may have states that qualify as representations in virtue of its very dynamics. For example, it follows from this definition that the angle of the arms of the Watt governor represents the speed of the engine. The angle of the arms stands between a representation producer, i.e. the flywheel, and a representation consumer, i.e. the throttle valve (R_1). In virtue of Watt's design, the function of the angle of the arms is to adapt the valve to the speed of the engine (R_2). Finally, there is a systematic correspondence between possible values of the arms' angle and the value of speed of the engine (R_3). So at the *metaphysical* level, it is true to say that there *are* representations in a Watt governor, at least if we accept Chemero's definition.

However, this does not entail that the representations instantiated in a dynamical system in virtue of its dynamics are *explanatory* of the behaviour of the system. I can extract three lines of argument from Chemero's discussion of the Epistemological Claim (2009: 67–84) that representations are explanatorily dispensable. First, dynamical modelling is explanatory in itself because it provides counterfactual-supporting generalizations. To stick with our toy example, the equations governing the Watt governor offer counterfactual-supporting generalizations of its behaviour: if the angle of the arms had been different, the speed of the engine would have been different. Second, representations are irrelevant to the explanatory character of the dynamical modelling because the dynamical model would remain explanatory even if the system instantiated no representations. Reasoning *as if* a system instantiated no representations is adopting what Chemero call the "dynamical stance" towards the system, consisting in

> explaining their behaviour with the tools of dynamical systems theory and avoiding representational vocabulary, while remaining agnostic on the status of the metaphysical claim. (2009: 72)

The "dynamical stance", is a term of art crafted by Chemero, and echoing Dennett's "intentional stance" (Dennett 1987). Both are "stances" in the sense that they consist in viewing a system *as if* it were X, while remaining non-committal as to whether it is really X. There is however an important disanalogy between the two: whereas Dennett's intentional stance is to look at systems as if they *had* beliefs and other intentional states while remaining agnostic regarding the existence of beliefs in that system, Chemero's dynamical stance asks us to do as if dynamical systems did *not* have any representations among their states. The second line of argument for the Epistemological Claim can then been summed up as follows: representations are epistemically dispensable because we incur no explanatory loss by adopting the *dynamical stance* towards the system under investigation. The use of representational vocabulary and the identification of explanatory factors with reference to their representational role does not provide better explanations than the ones available from the dynamical stance. Therefore, representations are epistemologically dispensable and the Epistemological Claim is true.

Chemero also uses a third line of argument to establish the Epistemological Claim, namely that the explanation offered by the dynamical modelling is *prior* to any explanation coined in terms of representation, in the following sense: assuming that the behaviour of the system has an alternative representational explanation, the representational explanation is not informative, given the dynamical explanation, whereas the dynamical explanation is still informative, even when one has already contemplated a representational explanation.

For example, one may argue, as Bechtel (1998, 2008) did, that the Watt governor was *designed* in such a way that the angle of its arms represent the speed of the engine and that this relation of representation explains why it succeeds in controlling the speed of the engine. Once this teleological-representational explanation of the success of the Watt governor is in place, the dynamical model explains additionally why this relation of representation holds and how it contributes to the success of the controlling task. However, once we know the equations of the dynamical system, we understand immediately why the Watt governor succeeds in its task, and the teleological-representational story has no additional explanatory value to offer. In this sense, the dynamical explanation of the behaviour of the governor is *prior* to the alternative teleological-representational explanation.

Let us note, in passing, that there is a tension between the notion of dynamical stance, as introduced in the quotation above, and the first point that dynamical explanations support counterfactuals. Let us ask ourselves what it is to consider this system as if it does not have any representation, in accordance with the dynamical stance. Obviously, when doing so, one still considers the system as a dynamical system, governed by a set of differential equations. But then, if these differential equations explain why some states of the system have representational properties, then representations will be instantiated in all the conterfactual situations where the dynamical equations hold. So one cannot coherently consider the system as lacking representations. If we look at the problem from another angle, considering the system as though it did not instantiate representations amounts to assuming that it is not governed by the set of dynamical equations that actually govern it. For keeping the equations fixed entails keeping the representations. Thus, for the kinds of dynamical systems which have representations, adopting the dynamical stance, as defined by Chemero, amounts to adopting an incoherent point of view on the system, considered both as instantiating and not instantiating representations. There is just no way to remain noncommittal or agnostic about the presence, *at the metaphysical level*, of representations in the system as long as one keeps the dynamics intact, if, by assumption, the dynamics explains the presence of representations.

Note also that this difficulty does not apply to Dennett's intentional stance, because the intentional stance consists in hypothesizing the additional existence of something, i.e. intentional states, whereas Chemero's dynamical stance consists in hypothesizing the absence of something, i.e. representations. Let us say that Dennett's intentional stance is an *additive* stance, whereas Chemero's dynamical stance is a *subtractive* stance. The problem identified above is a consequence of the *subtractive* nature of Chemero's dynamical stance. If the dynamical modelling of the Watt governor explains why some of its parts have representational properties, then it is incoherent to treat it as a system which does *not* have representational properties.

Perhaps what Chemero really wants to mean by "dynamical stance" is something slightly weaker, namely that one may describe a cognitive system as a dynamical system, without introducing any representational vocabulary and without identifying the components of the system by implicit reference to any representational relation toward some target in the environment. Intrinsic physical properties and dynamical roles should are the only bases one may rely upon to analyze a dynamical system. This is for sure a coherent stance to take towards cognitive systems. From now on, I will rely on this weaker characterization of the dynamical stance.

It might be objected—rightly, in my opinion—that the truth of the Epistemological Claim in the case of Watt's governor is a poor indicator of its truth for more sophisticated cognitive capacities where representations might be crucial. Chemero's argument in favor of the dynamical stance requires therefore to consider the relative merits of dynamical and representational explanations for uncontroversially *cognitive* capacities. He does indeed consider a few examples of more sophisticated cognitive agents, i.e. the "Sussex" robots (Harvey et al. 1997) and argues that adopting the dynamical stance allows to generate explanations of the robots' behaviour that are prior to any available representational explanation. Of course, whether the Sussex robots are sufficiently sophisticated to provide support to the epistemological claim as a general claim about cognition is open to discussion. A *prima facie* difficulty for the dynamical stance is to explain fully what Clark has dubbed "representation-hungry" tasks:

> Adaptive hookup phases gradually into genuine internal representation as the hookup's complexity and systematicity increase. At the far end of this continuum we find ... creatures that can deploy the inner codes in the total absence of their target environmental features. Such creatures are the most obvious representers of their world, and are the ones able to engage in complex imaginings, off-line reflection, and counterfactual reasoning. Problems that require such capacities for their solution are representation-hungry, in that they seem to cry out for the use of inner systemic features as stand-ins for external states of affairs. (Clarke 1997: 147)

Chemero offers a few examples of promising dynamical approaches to such capacities: Van Rooij, Bongers and Haselager's "non-representational approach to imagined action" (2002), i.e. imagining using a given stick to reach distant objects, and the study by Stephen, Dixon and Isenhower (2009) of gear-systems problem solving with the tools of dynamical systems theory. Still, it is still difficult to see how far this kind of approach can go, when it comes to other representation-hungry capacities. Time will tell how far the dynamical stance will prove to be explanatorily fruitful. But Chemero seems to be sufficiently optimistic about it to write a whole book defending a radical anti-representationalist program for cognitive science.

After this quick guided tour of REC, with a somewhat longer stop on the metaphysical and epistemological claims regarding mental representation, we are in a better position to evaluate this radical turn promoted by Chemero. Should we follow him and throw away "mental representations" from the vocabulary of embodied cognitive science?

There are several ways to resist such an elimination of mental representations from a dynamicist approach to cognition, some of which have been already explored in the literature. One way to resist the Radical move is to insist that there is no evidence that the success of non-representational dynamical explanations of elementary cognitive phenomena will project to more sophisticated, representation-hungry capacities (e.g. Clark and Toribio 1994). Another way is to attack the program on its own turf, so to speak, and question the claim that representations have no role to play in the explanation of those cognitive capacities that seem *prima facie* amenable to the dynamical stance (e.g. Bechtel 1998, 2008).

Reboul (2011) presents an original approach to this debate consisting in questioning the *substantiveness* of the debate over representation in the philosophy of cognitive science. Let us consider how this diagnosis applies to Chemero's specific proposal.

4 Begging the question about representation?

Reboul (2011) offers a diagnosis of the debate over the best way the understand the program of Embodied Cognition and in particular over the status of representations. The diagnosis is that much of the debate seems to be merely verbal.

When two philosophers or cognitive scientists argue about the necessity to posit representations to explain a certain type of behaviour, it may be useful for them to make sure that they understand the notion of representation in the same way. But there are several non-equivalent ways to approach the notion of mental representation. The sophisticated theories of mental representation proposed by philosophers of cognitive science not only differ in the way they explain how mental representations refer to their targets,[9] but also in the basic requirements they impose on the very concept of representation.

One of the crucial issues is that of *decouplability*, i.e. the possibility for a representation to represent its target in its absence, or at least without being causally connected to it. According to one view, influentially sketched by Haugeland, it is part of our most general notion of representation that a representation can be decoupled from its target:

> A sophisticated system (organism) designed (evolved) to maximize some end (such as survival) must in general adjust its behaviour to specific features, structures, or configurations of its environment in ways that could not have been fully prearranged in its design. If the relevant features are reliably present and manifest to the system (via some signal) whenever the adjustments must be made, then they need not be represented. Thus, plants that track the sun with their leaves needn't represent it or its position, because the tracking can be guided directly by the sun itself. But if the relevant features are not always present (manifest), then they can, at least in some cases, be represented; that is, something else can stand in for them, with the power to guide behaviour in their stead. That which stands in for something else in this way is a representation; that which it stands in for is its content; and its standing in for that content is representing it. (Haugeland 1998: 171)

Some philosophers share this view and have constructed a notion of representation which requires that the representation "stands in" for something distinct, and thus can serve its representational function in the absence of its target (van Gelder 1995, Smith 1996, Ramsey 2007).[10] With such a stringent notion of representation, it is not too difficult to argue that the explanations offered by the Embodied Cognition program are non-representational. For example, it is not too difficult to resist the claim that some part of the Watt governor represents the speed of the locomotive (van Gelder 1995). This relation of representation holds in virtue of a coupling between the governor and the engine. As soon

---

[9] See for example Stich and Warfield (1994) for a collection of classic papers.

[10] Apparently Haugeland later distanced himself from this view. See Chemero (2009: 213).

as the governor is decoupled from the engine, it cannot represent its speed any more.

The view that decouplability is a necessary condition of representationality has found some dissenters though. Consider for example the case of a group of neurons in the posterior parietal cortex of rats that respond maximally (in the context of running a radial maze) to specific combinations of head orientation and the presence of some local landmark or feature (McNaughton & Nadel 1990: 49-50).[11] It may seem excessive to deny that those neurons *represent* this type of combination just because those neurons could not play their role in the absence of a constant stream of proprioceptive signals from the rat's body in the presence of those features. For reasons like this one, another group of philosophers in the debate prefer to work with a more liberal notion of representation, according to which decouplability is not required (Clark 1997, Bechtel 1998, 2001, 2008). It is then easier to claim that the Watt governor does represent the speed of the engine (Bechtel 1998) and, more generally, that the explanations offered by the embodied cognition program are representational. So it looks like the disagreement between both groups of philosophers can be explained, at least for a good part, as a verbal dispute: they simply mean different things by "representation". And then each party in the debate can accuse their opponent of begging the question by adopting a definition of "representation" that favors their argumentative purpose.

Chemero is well aware of this difficulty. His solution is to avoid the accusation of begging the question by accepting the more liberal notion of representation that *prima facie* fits the needs of the representationalist side, while arguing for their explanatory dispensability. One might think that by accepting a definition of "representation" which *prima facie* fits the argumentative needs of his declared opponent rather than his own, Chemero avoids the charge of begging the question. Yet, it turns out that the choice of this liberal notion of representation, which does not require decouplability, and the focus on cases where the representation and its target are tightly coupled, Chemero can be accused of symmetrically begging the question against the view that representations are indispensable for the explanation of cognition: whenever the adaptive behaviour of an organism (or a robot) is successful in virtue of a tight coupling between the environment and the body, as described within a single nonlinear dynamical system, it is not too difficult to argue that the representational states of the organism are idle mediations between the body and the environment, and dispensable without much explanatory loss. Had one taken decouplability to be part of the definition of representations, one would have needed to show

---

[11] This example is borrowed from Clark (1997: 144).

that decouplable representations are epistemically dispensable. But, one could not argue for this claim by shewing examples where those representations are idle mediations because of a coupling with their targets. So the defence of the Epistemological Claim in this context seems more difficult, or at least would have to be quite different.

Even though Chemero claims that the tools of dynamical systems theory can describe systems capable of decouplable representations and explain some "representation-hungry" capacities, it is still not clear how these tools can provide interesting explanations of such cognitive capacities as the production and understanding of fiction, or the use of thought experiments, that require decouplable representations (Reboul 2011). So the risk, diagnosed by Reboul, that a major part of the debate regarding the possibility for the Embodied Cognition program to do away with mental representations stems from verbal disputes about the meaning of "mental representation" is still pressing, and measures to the contrary taken by Chemero are not wholly satisfactory.

Note also that this line of attack ultimately converges towards the first sort of criticism outlined above, i.e. the doubt that REC proves adequate when applied to sophisticated cognitive behaviour. However, this is not the only way to challenge REC and probably not the strongest. A stronger argument would be to show that the Epistemological Claim fails even for cognitive capacities that do not seem particularly "representation-hungry". This is the way I want to challenge REC in the remaining sections of this chapter. To do so, music cognition, and in particular rhythm cognition is arguably an interesting area to explore.

There are two reasons for this. First, rhythm cognition involves phenomena which *prima facie* fall under the purview of the Embodied Cognition Program. Our relationship with rhythm in music involves intimately our bodies (Leman 2008). We often move our bodies in response to musical rhythm. These movements, in dancing, or just tapping with a finger or a foot, seem to be an intrinsic part of the act of listening and understanding rhythms (Maes et al. 2014). So if anything should receive an illuminating explanation within the Embodied Cognition Program (Leman and Maes 2014), and *a fortiori* within the Radical Embodied Cognition Program—if Chemero is right—rhythm cognition should be part of it.

The second reason is more closely tied to an interesting detail in Chemero's discussion of the notion of representation. As we saw, he accepts that dynamical systems may instantiate representations. He also proposes to map classes of mental representation with classes of dynamical systems, in particular coupled oscillators. Decouplable mental representations are thus mapped to adaptive oscillators while non-decouplable ones are mapped to relaxation oscillators (Chemero 2009; Chemero & Eck 1999). And one of the specific examples he takes are oscillators used to model rhythm perception (McAuley 1995). So let us look at the Embodied Cognitive Science of rhythm and see if it supports Chemero's Epistemological Claim about representation as successfully as one might expect.

5 Rhythm perception and metric structure

In this section I consider some phenomena associated with the perception of rhythm structure. I would like to show how a REC theorist could make a case for the elimination of "mental gymnastics" from a good explanation of the perception of rhythmic structure. One way to do so is to contrast it with a more traditional representational-computational approach.

Here is how Lerdahl and Jackendoff (1983) characterize the phenomena of musical structure :

> One commonly speaks of musical structure for which there is no direct correlate in the score or in the sound waves produced in performance. One speaks of music as segmented into units of all sizes, of patterns of strong and weak beats, of thematic relationships, of pitches as ornamental or structurally important, of tension and repose, and so forth. Insofar as one wishes to ascribe some sort of "reality" to these kinds of structure, one must ultimately treat them as mental products imposed on or inferred from the physical signal. In our view, the central task of music theory should be to explicate this mentally produced organization. Seen in this way, music theory takes a place among traditional areas of cognitive psychology such as theories of vision and language. (1983: 2)

On this view, which is quite standard from the standpoint of classical cognitive science, one is said to understand a piece of music only if one is able to assign a structure to it. A crucial aspect of these phenomena is that the structure assigned to musical signals is in general *richer* that the structure of the physical signal itself.

For example, it is known since the pioneering work of Bolton (1884) on subjective rhythmization that isochronous sequences of monotone clicks of equal amplitude produce on hearers the remarkable impression that some

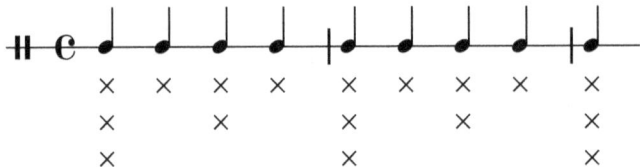

Figure 12.2: Metrical hierarchy associated to an isochronous beat

clicks are louder than others. Furthermore, the louder clicks recur periodically, in the sense that for some n, every $n + 1^{th}$ click is heard as louder than the preceding $n - 1$ clicks. The value of n can vary with the tempo of the isochronous sequence and other factors. The values of n can range between 2 and 8, but the most common values are $n = 2, 3, 4$.[12]

Lerdahl and Jackendoff, in their 1983 book, provide a theory of metrical structure which purports to account for these phenomena and more sophisticated ones. Let me first introduce some basic concepts of rhythm structure. *Beats* are isochronous idealized events of infinitesimal duration. A series of beats forms a *metrical structure* only if there is a distinction between strong and weak beats. For beats to be strong or weak there must exist a metrical hierarchy, that is two or more levels of beats. In 4/4 meter, for example, the first and third beats are felt to be stronger than the second and fourth beats and so are represented at the next larger level. The first beat is felt to be stronger than the third beat and so appears at the next level of metric organization, and so on (see Figure 12.2 for the simple case of an isochronous beat).

The basic question, then is how metrical structures, so defined, are to be assigned to musical signals. To answer this question Lerdahl and Jackendoff provide a series of rules. They distinguish in particular two kinds of rules, well-formedness rules (WFRs) and preference rules (PRs). Rules of the first kind define the set of possible metric structure in the idiom under investigation. The two authors give four basic WFRs, two of which are claimed to be universal, the other two being peculiar to the tonal idiom they are specifically interested in. A universal WFR is for example the rule that every beat at a given level should also be a beat at the smaller levels present at that point in the piece. The rule that in each metrical level, strong beats are spaced either two or three beats apart, unlike the preceding, is characteristic of classical tonal music.

The PRs then address the problem of relating these possible metrical structures to a given musical surface. For it can happen that various possible metrical structures are compatible with a same musical surface—and it does happen

---

[12] See Bååth (2015) for a recent study of subjective rhythmization.

Figure 12.3: According to the Event Preference Rule, the metrical structure (a) is preferred to the metrical structure (b).

a lot when we consider sophisticated pieces of music. The task of the PRs is then to specify the metrical structure that corresponds to the musical intuitions of the listener in the idiom under consideration. These preference rules are usually connected to other aspects of rhythm structure, for example grouping structure (describing how notes are heard as forming a rhythmical unit) or dynamic structure (describing the relative loudness or softness of notes within a group). For example, the Event Preference Rule asks us to prefer a metrical structure in which beats of level L_i that coincide with the inception of pitch-events are strong beats of L_i (see Figure 12.3).

Lerdahl and Jackendoff's tour de force is to have provided a system of rules capable of explaining the way we hear relatively complex pieces of music, among which some of the masterpieces of European classical music. Figure 12.4 is an example of an assignment of metrical structure to the beginning of Mozart's Symphony in G minor. Each note in the melody is assigned a relative strenght, corresponding to the level it occupies in the metrical hierarchy. A note on the fourth level, e.g. the D on the second measure, will "sound" rhythmically stronger than a note on the third level, e.g. the G at the beginning of the third measure.

It has to be noted that the theory offered by Lerdahl and Jackendoff intends to capture the tonal and rythmical *competence* of hearers. Their goal can be described as specifying a function which assigns to any musical signal a structural description correspond to the way it is heard by a normal hearer, in the same way that syntactic theory is concerned with the assignment of a syntactical description to sentences in a given natural language. If we restrict our attention to rhythm, an important part of the theoretical goal is to predict

Figure 12.4: Metrical structure of the opening theme of Mozart's Symphony No. 40 in G minor, KV. 550.

the metrical structure which corresponds to the way the music is heard by a normal competent hearer.

There is a sense, then, in which the theory they offered is psychologically incomplete:

> Instead of describing the listener's real-time mental processes, we will be concerned only with the final state of his understanding. (Lerdahl & Jackendoff 1983: 3)

Because their primary purpose was to set out a theory of the *competence* of a hearer familiar with the tonal idiom, they did not inquire into the nature and details of the real-time mechanisms at work in rhythm perception, among other aspects of music perception. For example, the preference rules are not supposed to describe the mechanism by which a metrical structure is assigned to a given rhythmical stimulus, as it is processed, beat after beat. This rules are only part of a system which describes what the mechanisms *should arrive at*. Thus, given a musical surface, an account of the mechanisms of rhythm perception should specify the effective procedure by which the mind constructs in real time a mental representation of the rhythm structure from the acoustic input. Still, their system of rules is an important achievement that puts a substantive constraint on the details of those mechanisms.

The general view exposed in the quotation above, and the specific rule-based theory of rhythm perception just outlined suggest that Lerdahl and Jackendoff envisage music cognition as a sort of "mental gymnastics" as Chemero would put it. The mind somehow applies a set of rules to given representations in order to process new representations. The main argument for this sort of approach seems to be an argument from the poverty of the stimulus: the structure we assign to the musical surface is in general much richer than the structure present in the musical signal ; *therefore* it must be imposed by the mind. This line of reasoning might seem reasonable, at least *prima facie*. If the structure is not in the signal, it has to be generated somehow endogenously. But this does not entail that the endogenous mechanism responsible for this has to be mental. So the mentalistic aspect of the conclusion of the argument one may attribute to Lerdahl and Jackendoff seems unwarranted.

In fact, this is precisely what some recent research, relying on the tools of dynamical systems theory cast into doubt. As it turns out, rhythm perception is a field where the introduction of the sort of tools promoted by REC have offered interesting results.[13] In particular, the basic idea of Neural Resonance Theory (Large 2008; Large & Snyder 2009; Large, Herrera & Velasco 2015) is to model the perception of metric structure by a network of coupled nonlinear oscillators. When this network receives a rhythmic signal as an input, the network creates a dynamical pattern of oscillation which synchronizes at various periodicities with the incoming rhythm.

This theoretical approach is based on the concept of neural oscillation. Individual neurons and populations of neurons are known to generate oscillatory activity in many ways. Neural Resonance Theory (NRT) hypothesizes that some patterns of neural oscillation, modelled by appropriate networks of coupled oscillators, account for various aspects of the perception of rhythmic structure, and metrical structure in particular. Unlike biophysical models of neural oscillation, where each individual neuron is modelled by its own set of equations, oscillator models are supposed to capture more abstract patterns of oscillatory activity. They are simplified but mathematically more tractable models of neural oscillation. One can also consider the canonical model of an oscillator model, which abstracts away from particular parameters, and represents common properties of a class of dynamically equivalent oscillator models.

Among the common properties exhibited by the canonical model of the Wilson-Cowan oscillators, one finds three important features.[14] First, oscillators have a capacity for *spontaneous oscillation*, in the sense that they can have oscillatory behaviour in the absence of input. Second, oscillators have a capacity for *entrainment*: when a stimulus is present, the spontaneous oscillation continues, but the coupling with the stimulus affects the phase of the

---

[13] See Levitin et al. (2018) for a general review of recent work in rhythm cognition and Large (2010) for a slightly older overview of the contribution of dynamical systems theory to music cognition. More recent work will be cited in the coming pages.

[14] The following description of the properties of this model follow the presentation of (Large 2008) where the reader will find the mathematical details.

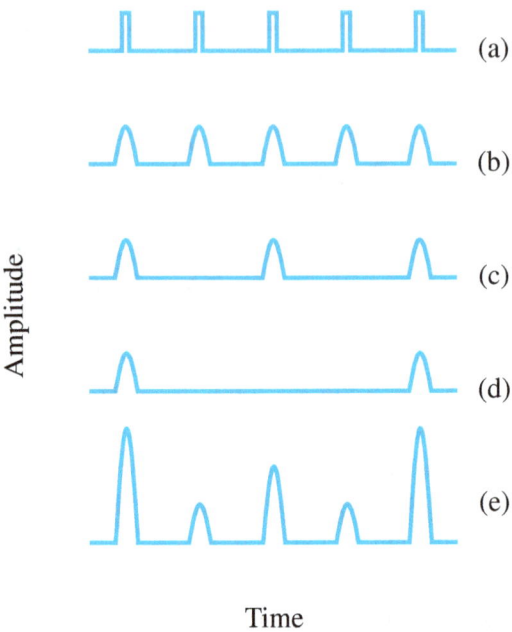

Figure 12.5: Schematic plot of subjective rhythmisation in a resonance theory framework. Adapted from Bååth R. (2015: 246). (a) represents the sound input, (b), (c) and (d) represent three individual oscillators while (e) represents the whole network of oscillators.

oscillator, so that the oscillator can synchronize with the stimulus at various ratios (e.g. 1:1, 1:2, 2:1, 1:3, 2:3). Third, networks of oscillators have a capacity for *higher-order resonance*. When a network of oscillators, each having a different frequency within an appropriate range, is stimulated with a periodic signal, various oscillators in the network will synchronize at various ratios of frequency of the periodic input signal.

These general properties of neural oscillations predicted by the canonical model can be connected at the phenomenological level with various aspects of rhythm cognition and seem to provide the basic ingredients for their explanation: *spontaneous oscillation* may account for the experience of endogenous periodicity; *entrainment* may explain our capacity to coordinate with rhythms at various levels of periodicity; and *higher-order resonance* may explain the metrical hierarchy assigned to pulsed rhythmic stimuli (see Figure 12.5 for the simple case of a isochronous beat)

Each of these hypotheses have been explored and confirmed to a certain extent (Large 2008; Large, Herrera & Velasco 2015). In particular, NRT has been shown to explain phenomena associated with subjective rhythmization (Bååth 2015) and, remarkably, to assign metrical structures to non trivial syncopated rhythms (Large, Herrera & Velasco 2015; Tal et al. 2017).

The way we presented NRT focused on the coupling between neural oscillations and the physical signal produced by a musical stimuli. This may suggest that the theory focuses mostly on the brain, and does not have anything particularly interesting to contribute to the program of *Embodied* Cognition. Jumping to such a conclusion would be wrong, however. In some recent presentations of the theory (Tal et al. 2017), the proponents of NRT explicitly distinguish two oscillatory networks, one representing the physical properties of the stimulus, and a second one that integrates inputs from the motor system. This enables the theory to integrate experimental data showing, for example, that some ambiguous rhythms can be disambiguated with the help of bodily movements (Philips-Silver & Trainor 2005, 2007, 2008; Trainor & Gao 2009).

So NRT is a theoretical framework within which a number of phenomena related to Embodied Music Cognition can be studied and to a certain extent explained. It should be pointed out that this approach faces a number of challenges, regarding in particular the empirical investigation of its neural basis. The models provided so far are very abstract and some more work is still needed to connect the mathematical models to the neurobiological reality (see however Large E.W., J.A. Herrera & Velasco M.J. 2015 and Tal et al. for two recent attempts). Moreover, NRT is by no means the only game in town. Some important criticism have been voiced and alternative approaches are being explored.[15] This being said, it still appears as an important attempt at building, with the tools of Dynamical Systems Theory, a bridge between neural oscillations and rhythm cognition.

Let us suppose, at least for the sake of the argument, that NRT is on the right track. Then it allows to explain the assignment of a metrical structure to a musical signal without introducing any "mental gymnastics". As Large puts it:

> ...pulse and meter are seen not as computational "problems" to be solved by the brain; they are simply what happens when nonlinear resonators, operating at the proper timescale, are stimulated by music. ...They are intrinsic to the physics of neural oscillation. (Large 2008: 233)

So we have, *prima facie*, a very good example of the way a dynamical ap-

---

[15] See for example Patel and Iversen's Action Simulation for Auditory Perception Hypothesis (Patel & Iversen 2014).

proach can be successful without positing computations over representations. It seems one can explain the assignment of metrical structure from the dynamical stance alone. So this example apparently provides additional support to Chemero's Epistemological Claim.

6 Rhythm and Representation

The devil, as always, is in the details. Let us consider more precisely how Chemero's argument for the Epistemological Claim would apply here.

First, it should be clear that networks of oscillators do instantiate representations at the metaphysical level. Chemero suggests as much when he proposes to map classes of representations to classes of oscillators. It is easy to confirm this claim by applying the criteria of representation (R1)-(R3) to the present case. For example, one can say that a resonating oscillator (see Figure 12.5) represents an aspect of a rhythmic stimulus, namely a level of its metrical structure. In accordance with (R1), the network of coupled oscillators is a representation producer and the motor system is a representation user whenever the agent relies on her understanding of the metrical structure of the rhythm in order to synchronize bodily movements with it. In accordance with (R2), it is the function of the oscillator to represent this aspect of the rhythmic stimulus, but it could of course go wrong for a number a reasons—if the tempo is very high or very low, of if the rhythm is subtly ambiguous. Finally, in accordance with (R3), there is a systematic relation between changes in the frequency of that oscillator and a successful adaptation to e.g. the tempo of the rhythmic stimulus, for the representation consumer: if the tempo of the rhythm slows down, the oscillator will have to slow down to remain synchronized with it.

So it is possible to provide a representational description of the the neural oscillations modelled dynamically by a network of coupled oscillators. For Chemero's Epistemological Claim to hold, however, adopting a purely dynamical stance towards this system should not lead to any explanatory loss and the dynamical explanation should be prior to any alternative representational explanation.

The difficulty here is that it seems impossible to remove all representational vocabulary from the dynamical explanation of rhythm perception offered by NRT. First, the *explanandum* essentially involves a representation. What we want to explain is how a metrical hierarchy is ascribed to a rhythmic stimulus. By definition, the metrical hierarchy is not a physical property of the rhythmic stimulus, on the same footing, for example, as the amplitude of the signal, or the onset times of the notes composing the rhythm. The metrical hierarchy exists insofar as it is represented by the hearer. Therefore, explaining the assignment of metrical structure implies the ascription of representations, otherwise it is hard to see how the *explanans* connects to the *explanandum*. If one were to adopt the dynamical stance and remove all explicit or implicit reference to representation, then it would be impossible for the dynamical description to capture the phenomenon it is meant to explain. The metrical structure, as represented by the hearer, would simply vanish. The only thing that remains is a physical signal devoid of any metrical structure.

The present point may be better appreciated with the help of a comparison with other examples of cognitive phenomena that Chemero takes to support the Epistemological Claim. In the case of Beer's artificial agent or the Sussex robots the *explanandum* is not defined directly in terms of representing a target in a certain way, but rather in terms of successfully achieving a particular task describable without any representational vocabulary (e.g. moving towards a potentially moving target, grasping a certain type of object, avoiding another type of object). The gist of the argument for the Epistemological Claim, in these cases, is that positing representations is unnecessary to explain this success, as a dynamical explanation is already available and independently satisfactory. Here, the situation is different in that the *explanandum* itself involves a representation. Here a non-representational dynamical story has no chance to be explanatory because it will not be in a position to even capture adequately its *explanandum* and *a fortiori* to explain it. To sum up, if one were to adopt the dynamical stance, the *explanandum* would vanish and no explanation would be available.

The second point concerns the *explanans* rather than the *explanandum*. It is quite clear that the components of the dynamical systems described by NRT are individuated with implicit reference to their representational role. The reason why such a multiplicity of oscillators are posited in the model is precisely because one wants to model various levels of synchronization, corresponding to various levels of metrical structure. So one does not adopt the dynamical stance when one models rhythm perception with networks of coupled oscillators in the NRT does. Under the dynamical stance, the components of the system are not individuated with reference to their representational role, but only with reference to their dynamical role and their intrinsic physical properties. But it seems clear that the structure of the model is itself based on the goal to model metrical structure rather than on an independent hypothesis about the intrinsic properties of neural oscillations. It may be interesting in this regard to note that NRT was first conceived (e.g. Large 2002) as an abstract mathematical description of rhythm perception, without any claim of neural plausiblity.

It is only later that a connection with *neural* oscillation was hypothesized at the theoretical level, and tested at the empirical level. Even though we have independent evidence of the existence of neural oscillations at various periodicities, the idea that some specific patterns of oscillations may account for metrical structure does not come from independent neurobiological grounds. It is rather that some appropriate mathematical models may turn out to be neurobiologically plausible. This confirms the view that the analysis of the models used in NRT make implicit reference to a represented metrical structure, and therefore do not belong strictly speaking to the dynamical stance.

A proponent of REC may retort that the initial characterization of the *explanandum* in terms of representation begs the question against her view. A truly *Radical* Embodied Cognition theorist may want to resist this characterization. Instead, one should focus on the behaviour of the whole dynamical system comprising the brain, the body and the musical stimulus in the environment. The target of the explanation, then, should be the *synchronization* of motor movements to the rhythm, e.g. tapping, at various periodicities, rather than the *representation* of a metrical hierarchy.

It is not easy to make that characterization fully precise, though. First, in most cases, a metrical hierarchy is represented in the absence of any motor output, so the REC theorist should insist that the target of the explanation is a mere *disposition* to synchronize some motor behaviour at various periodicities with a rhythmic stimulus. For example, in the case of the Mozartian theme considered above, the target of the explanation is a disposition to tap spontaneously according to patterns in Figure 12.4. More precisely:

- if one were to tap at the frequency of eight notes, one would tap all half beats;

- if one were to tap at the frequency of quarter notes, one would tap all beats

- if one were to tap at the frequency of half notes, one would tap only those beats marked at the third level of metrical structure;

- if one were to tap at the frequency of whole notes, one would tap the beats marked at the fourth level of the metrical structure.

Second, even this more sophisticated rephrasing of the *explanandum* fails to capture the *hierarchical* element of metrical structure. And it is hard to see how to characterize it, except by the fact that the beats of higher levels are *felt*, i.e. *represented*, as stronger.

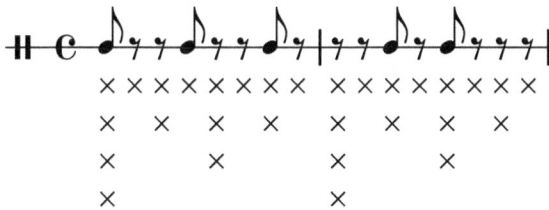

Figure 12.6: A syncopated rhythm with its intuitive metrical structure. The first beat of the second bar occurs at the strongest level although it is occupied by a rest.

But let us suppose, for the sake of the argument, that such an appropriate non-representational description of the *explanandum* can be provided. Still, this Radical view would be at odds with the explanatory practices of the researchers working on Embodied Music Cognition within the framework of dynamical systems theory. A significant bulk of the work done in that area focuses on showing that the higher-order resonance patterns of coupled oscillators match the *intuitive* metrical structure associated with particular rhythmical stimuli. By "intuitive metrical structure", I mean the structure (assumed to be) assigned by a normal competent hearer, rather than the metrical structure revealed by the motor behavior of a subject.[16]

In fact, it is crucial for the empirical testing of NRT that neural oscillations observed by MEG be compared to *intuitive* metrical structure, rather than to behavioral indications that a beat is actually felt as stronger by the subject. For in the latter case, it would be difficult to separate in the data what is caused by the mere perception of rhythm from what is caused by the act of tapping or moving with the rhythm. For this reason, the experimental research has focused on the study of syncopated rhythm (Tal et al. 2017) where the beats occurring at the highest level of the metrical structure do not even have a physical counterpart in the musical stimulus (see Figure 12.6). If oscillations are recorded at this frequency in the absence of tapping, then those possible confounds are neutralized and NRT gets a clearer confirmation. The upshot of this example, is that reference to a represented metrical structure is essential to the current explanatory practice of the field. So the REC program would require a drastic revision of the current explanatory practices.

This is not necessarily a problem in itself. After all, the point of REC is to

---

[16]Of course, this requires the assumption that the subjects's perception of the rhythm conforms to that of such a normal hearer. This assumption can be, and is usually, controlled for with the help of an independent tapping task (see e.g. Tal et al. 2017).

call for a change in the explanatory practices of cognitive science. However, this commitment to a revision of the explanatory practices relying on *dynamical systems theory* raises a dialectical difficulty for the proponent of REC. The kind of arguments that are used to support REC, in Chemero's book at least, consist in showing that dynamical explanations, as they are *de facto* offered, validate the Epistemological Claim. But if the validation of the Epistemological Claim requires a reinterpretation or a revision of the actual dynamical explanations, the argument for REC is less compelling, for it is not based on clear examples of successful non-representational dynamical explanations, but rather on *suitably reinterpreted* examples. I thus conclude that the dynamical explanations of meter perception promised by NRT do not support the Epistemological Claim. Rather they suggest that the use of dynamical systems theory in cognitive science do not necessarily entail the radical rejection of representations.

7 Conclusion

Chemero (2009) has proposed a Radical construal of the Embodied Cognition Programme according to which the old-fashioned computational-representational explanations of classical cognitive science should be abandoned in favour of dynamical explanations. On this view, representations cease to play any fundamental explanatory role and, for that reason, should be dispensed with altogether.

I have explored two ways to challenge this proposal. One follows a diagnosis offered by Reboul (2011), according to which the dispute between representationalists and anti-representationalists threatens to be merely verbal. Chemero is in fact aware of this difficulty, but I argued that his solution is not fully satisfactory. The other consists in identifying cases of successful dynamical explanations where representations cannot be eliminated because they are presupposed by the very phenomenon to be explained and because the components of the system are individuated with reference their representational roles. The phenomena of rhythm perception I considered do not involve decouplable representations, nor are they associated with "representation-hungry" tasks. For these reasons, they raise a difficulty from inside the area where REC seems to be the most promising.

So the ban of representation is perhaps not a welcome gift in the REC package, even if one acknowledges the fruitfulness of dynamical systems theory. The illuminating character of dynamical explanations appears to be independent from the explanatory status of representations. In some cases, representations may be explanatorily idle, in others they are essential. So promoting the use of dynamical tools and campaigning against representations in cognitive science are two distinct businesses, only the first of which seems ultimately profitable to the program of Embodied Cognition.[17]

7.1 References

Bååth R. (2015). "Subjective Rhythmization. A Replication and Assessment of two Theoretical Explanations". *Music Perception* 33(2): 244–254.

Bechtel W. (1998). "Representations and cognitive explanations". *Cognitive Science* 22: 295–318.

— (2001). "Representations: from neural systems to cognitive systems". In W. Bechtel, P. Mandik, J. Mundale & R. Sufflebeam (eds). *Philosophy and the Neurosciences*. Oxford: Blackwell Publishing. 332–348.

— (2008). *Mental Mechanisms*. New York: Routledge.

Beltrami E. (1987). *Mathematics for Dynamic Modeling*. Boston: Academic Press.

Bolton T. (1894). "Rhythm". *The American Journal of Psychology* 6(2): 145–238.

Chemero A. (2009), *Radical Embodied Cognitive Science*, Cambridge MA: MIT Press.

Chemero A. & D. Eck (1999). "An exploration of representational complexity via coupled oscillator systems". In U. Priss (ed). *Proceedings of the 1999 Midwest AI and Cognitive Science Conference*. Cambridge: AAAI Press.

Clark A. (1997) *Being There: Putting Brain, Body and World Together Again*. Cambridge: MIT Press.

— (2008). *Supersizing the Mind: Embodiment, Action and Cognitive Extension*. Oxford: Oxford University Press.

Clark A. & J. Toribio (1994). "Doing without representing?". *Synthese* 101: 401–431.

Dennett D.C. (1987). *The intentional stance*. Cambridge MA: MIT Press.

---

[17]A distant ancestor of this chapter was given as a presentation in a workshop around the themes of Shaun Gallagher at the École Normale Supérieure de Lyon in the presence of Shaun Gallagher, Jean-Michel Roy, Leon de Bruin and Annika Fiebisch, all of whom are thanked for valuable feedback. The last version benefited from discussions with Andrea Ravignani and Manon Grube on the neuroscience of rhythm perception—any imprecision in that area is attributable to the author only—and from the feedback of Clément Canonne. The writing of this paper was supported by the project Intuitions in Science and Philosophy funded by the Danish Council for Independent Research DFF 4180-00071.

Harvey I., P. Husbands, D. Cliff, A. Thompson & N. Jakobi (1997). "Evolutionary robotics: The Sussex approach". *Robotics and Autonomous Systems* 20: 205–224.

Large E.W. (2008). "Resonating to Musical Rhythm: Theory and Experiment". In S. Grondin (ed). *The Psychology of Time*. Cambridge: Emerald. 189–231.

— (2010). "Neurodynamics of Music". in M.R. Jones et al. (eds). Music Perception. New York: Springer. 201–231.

Large E.W., J.A. Herrera & Velasco M.J. (2015). "Neural Networks for Beat Perception in Musical Rhythm." *Frontiers in Systems Neuroscience* 9 (159).

Large E.W. & J.S. Snyder (2009). "Pulse and Meter as Neural Resonance." *Annals of the New York Academy of Science* 1169: 46–57.

Leman M. & P.J. Maes (2014), "Music perception and Embodied Cognition", In Shapiro L. (ed). *Routledge Handbook of Embodied Cognition*. London and New York: Routledge. 81–89.

Lerdahl F. and R. Jackendoff (1983) A Generative Theory of Tonal Music, Cambridge MA, The MIT Press.

Levitin D.L., Grahn J.A. and London J. (2018). "The psychology of Music: Rhythm and Movement". *Annual review of Psychology* 69: 51–75.

Maes P.J., M. Leman, C. Palmer & M.M. Wanderley (2014). "Action-based effects on music perception". *Frontiers of Psychology* 3 (4): 1008.

McAuley J.D. (1995). *Perception of Time as Phase: Toward an Adaptive-Oscillator Model of Rhythmic Pattern Processing*. PhD Dissertation, Computer Science and Cognitive Science, University of Indiana.

McNaughton B., & L. Nadel (1990). "Hebb-Marr Networks and the neurobiological representation of action in space". In M. Gluck & D. Rumelhart (eds). *Neuroscience and Connectionist Theory*. Hillsdale: Erlbaum.

Millikan R.G. (1984). *Language, Thought, and Other Biological Categories*. Cambridge MA: MIT Press.

— (1993). *White Queen Psychology and Other Essays for Alice*. Cambridge MA: MIT Press.

Ramsey W. (2007). *Representation Reconsidered*. Cambridge: Cambridge University Press.

Reboul A. (2011). "Radical embodied cognition vs. 'Classical' embodied neuroscience". *The Journal of East China Normal University*, 6.

Routledge R. (1900). *Discoveries and inventions of the nineteenth century*. London: Routledge & Sons. 13th edition.

Shapiro L. (2010). *Embodied Cognition*. London & New York: Routledge.

Shapiro L. (ed). (2014). *The Routledge Handbook of Embodied Cognition*. London: Routledge.
Smith B.C. (1996). *On the origins of objects*. Cambridge MA: MIT Press.
Stich S.P. & Warfield T.A. (1994). *Mental Representation. A Reader*. Oxford: Blackwell.
Strogatz S.H. (1994) *Nonlinear Dynamics and Chaos*. Cambridge MA: Westview Press.
Pain H.J. (2005). *The Physics of Vibrations and Waves*. Chichester: Wiley.
Phillips-Silver J. & L. Trainor (2005). "Feeling the Beat: Movement Influences Infant Rhythm Perception". *Science* 308: 1430.
— (2007). "Hearing what the body feels: Auditory encoding and rhythmic movement". *Cognition* 105: 533–546.
— (2008). "Vestibular influence on auditory metrical interpretation". *Brain and Cognition* 67: 94–102.
Pylyshyn Z. (1984). *Computation and Cognition*. Cambridge, MA: MIT Press.
Trainor L. J. & X. Gao (2009). "The primal role of the vestibular system in determining musical rhythm" *Cortex* 45: 35–43.
Stephen D., J. Dixon & R. Isenhower (2009). "Dynamics of representational change: Entropy, action, cognition". *Journal of Experimental Psychology: Human Perception and Performance* 35(6): 1811–1832.
Tal I., E.W. Large E.W., E. Rabinovitch, Y. Wei, C.E. Schroeder, D. Poeppel & E. Zion Golumbic (2017). "Neural Entrainment to the Beat: The "Missing-Pulse" Phenomenon." *The Journal of Neuroscience* 37(26): 6331–6341.
Thelen E. & L. Smith (1994). *A Dynamic Systems Approach to the Development of Cognition and Action*. Cambridge MA: MIT Press.
van Gelder T. (1995). "What might cognition be, if not computation?". *The Journal of Philosophy* 92(7): 345–381.
van Rooij I., R. Bongers & W. Haselager (2002). "A non-representational approach to imagined action". *Cognitive Science* 26: 345–375.
Varela F., E. Thomson & E. Rosch (1991). *The Embodied Mind: Cognitive Science and Human Experience*. Cambridge MA: MIT Press.

Chapter 13
Preschoolers' Social Preferences in a Dominance Context

Rawan Charafeddine, Chloé Billamboz,
Ira Noveck & Jean-Baptiste Van der Henst

Abstract Navigating the social world requires evaluating how others behave, communicate and feel. Children show signs of social evaluation from an early age by preferring those who are familiar, those who are similar to themselves as well as those displaying moral behaviour. In the current study, we investigate whether another key dimension of the social environment—social dominance—influences preschoolers preferences. Research shows that preschoolers understand such relations and use them to make social inferences. Less is known about their preferences towards dominant and subordinate individuals. We carried out two experiments. Experiment 1 presented 4- and 5-year-old children with a dominance scenario in which one dominant character twice imposed his/her will on a subordinate. The results showed that the children did not reveal a preference for one character over the other. In Experiment 2, 3- to 5-year-old children were presented with more explicit dominance interactions involving puppets in a decision power scenario (similar to Experiment 1's) and in a play-fight scenario. In the decision power situation, only the 3-year-olds revealed a preference for the dominant; also, boys were more likely to prefer the dominant than girls. In the play-fight scenario a slight preference emerges for the puppet that prevailed in the fight.

Keywords Social dominance · Social preferences · Social cognition · Preschoolers · Children · Developmental psychology

1 Social evaluation in early ages

Affiliating with partners with whom interactions can be fruitful and avoiding hazardous individuals is essential for social species. That is, navigating the social world requires evaluating others' behaviours, intentions and feelings. Social evaluation enables one to assess the costs and benefits of potential short-term and long-term relationships, and enables one to identify individuals who are more cooperative or who occupy valuable positions in a group. The need for positive interactions emerges early in human ontogeny. Research shows that from a very young age children are capable of social evaluation. Preschoolers and even infants prefer individuals over others on the basis of various social criteria. A first factor that determines children's social preference is familiarity. Newborns prefer their mother's voice (DeCasper & Fifer 1980) and three- to four-month-old infants show a visual preference for faces of the same gender as their primary caregiver (Quinn, Yahr, Kuhn, Slater & Pascalis 2002). Moreover, at 5 to 6 months of age, infants look preferentially at a speaker of their native language over a speaker of a foreign language and when preschool children are explicitly asked to choose a friend between an individual who speaks their language and a foreign speaker, they select the former (Kinzler, Dupoux & Spelke 2007).

Another factor that drives children's social preference is similarity, which involves social categories but also behavioural cues. Preverbal infants like more puppets that hold the same objects or prefer the same food as themselves than puppets that do not (Mahajan & Wynn 2012, see also Fawcett & Markson 2010 for similar results with 3-year-old children). Between 2 and 3 years of age, children affiliate and play more with peers of the same gender and same age as themselves (Challman 1932; Chevaleva-Janovskaja 1927; La Freniere, Strayer & Gauthier 1984). In the same vein, when preschoolers are explicitly asked which children they like more or which children they would like to play with, they indicate same-age and same-gender peers (French 1984; Martin 1989; Martin, Fabes, Evans & Wyman 1999). Preschoolers also align their choices with those of same-gender and same-age as themselves (Shutts, Banaji & Spelke 2010).[1]

Research also shows that children's preferences are influenced by proso-

---

[1] Regarding race, children's preferences are less clear. Although 3-month old infants show a visual preference for own-race faces over other-race faces (Bar-Haim, Ziv, Lamy & Hodes 2006; Kelly et al. 2005), this bias disappears in an environment where other-race faces are more familiar than own-race faces (Bar-Haim, Ziv, Lamy & Hodes 2006). Moreover, explicit affiliative preferences for own-race peers emerge later for race than for same-gender and same-age peers (Kinzler & Spelke 2011; Shutts, Banaji & Spelke 2010).

cial and moral behaviour at a very early age. Six-month-olds show their preference by reaching for a character who helps another perform an action over a character who hinders another, and prefer to reach a neutral individual over a hinderer (Hamlin, Wynn & Bloom 2007). Looking time measures even show that the aversion to hinderers is already present at 3 months of age (Hamlin, Wynn & Bloom 2010). Moreover, at preschool ages, children distribute fewer resources to a hinderer than to a helper character (Kenward et Dahl 2011) and are less likely to help perpetrators of harmful acts (Vaish, Carpenter & Tomasello 2010).

2 The case of social dominance

Familiarity and similarity are signs of the social proximity among individuals in the environment while morality provides information about how they behave with others. These factors can inform a social agent about the degree to which others are likely to cooperate with him or her. However, another key dimension of the social world, which does not characterize the inclination to cooperate, is *verticality* (Hall, Coats & LeBeau 2005). This dimension distinguishes group members according to the advantages they benefit from and characterizes social inequalities or asymmetries in the group. In the current study, we investigate whether one manifestation of verticality, namely *social dominance*, influences preschoolers' preferences. Dominance hierarchies constitute a recurring feature that structures the social life of humans and other species. They typically result from competitive interactions and determine which individuals have priority of access to resources in a group. In humans, hierarchies are found in all societies (Boehm & Boehm 2009; Brown 1991), and social dominance structures spontaneously emerge in groups of preschoolers as young as two (Boyce 2004). Investing cognitive resources in the representation of others' rank in a dominance hierarchy enables one to understand the social dynamics of a group functioning as well as the outcome of dyadic interactions within the group. In turn, this may help a group member to decide who she should affiliate with (Byrne & Whiten 1988). Affiliating with and imitating dominant individuals can be a way to obtain the same advantages as those individuals and to enhance one's own social rank. It can also be a means to avoid costly interpersonal conflicts with those individuals. However, a dominant individual often acts in a selfish and antisocial manner with respect to the subordinate. Social creatures that are sensitive to other's moral behavior may thus dislike them.

Research has shown that children show sophisticated cognitive skills that allow them understanding hierarchy. First, they represent hierarchy on the basis of a variety of asymmetrical cues. Using looking time paradigms, studies have reported that infants take into account body size and numerical alliances to predict the outcome of a conflict between two individuals (Pun, Birch & Baron 2016; Thomsen, Frankenhuis, Ingold-Smith & Carey 2011). To make explicit judgments of dominance and power (e.g. "Who is the boss?", "Who is in charge?"), preschoolers use physical cues, such as facial and body postures, but also information about age and wealth (Brey & Shutts 2015; Charafeddine et al., 2015; Keating & Bai 1986).

Second, children's representations of dominance also rely on observable interactions. Fifteen-month infants expect the outcome of competitive interactions to be stable over time: if they observe an individual push another one to monopolize a small territory, they expect the former to prevail over the latter when they will compete for monopolizing an object (Mascaro & Csibra 2012). Preschoolers are also sensitive to non-agonistic interactions to decide who the boss is. Three- to 5-year-old children tend to consider that an individual who imposes on others what game to play or who denies permission to others to use resources is more likely to be the boss (Charafeddine et al. 2015; Gülgöz & Gelman 2017). Moreover, at five years of age, children assign greater power to an individual who is imitated or who sets norms for others (Gülgöz & Gelman 2017; Over & Carpenter 2015).

Third, another manifestation of children's ability to think about hierarchies lies in their inferential abilities. Infants (10-13 months) can make transitive inferences in the context of dominance relationships: if they observe that A prevails over B in resource competition, and B prevails over C, they expect A to prevail over C (Gazes, Hampton & Lourenco 2017). Children can also draw inferences from one dimension of hierarchy to another. Three- to 5-year-olds predict that a powerful character, who gives orders to another compliant character, will win a game against and will also have more resources than the compliant character (Charafeddine et al. 2015).

This latter result suggests that children associate high rank with positive traits but a positive evaluation of a dominant position arises more straightforwardly in tasks involving resource allocation, testimony assessment, and self-perception. Three- and four-year-old children allocate more resources to a dominant than to a subordinate puppet (Charafeddine et al. 2016) and, when confronted with two contradictory testimonies, preschoolers endorse the testimony of the dominant individual over that of the subordinate one (Bernard et al. 2016; Castelain, Bernard, Van der Henst & Mercier 2016). The valuing of dominance is also found when preschoolers perceive themselves as dominant rather than subordinate and when they overestimate their dominance status in their peer group (Charafeddine et al. in press; Edelman & Omark 1973; Sluckin & Smith 1977). However, it would be hasty to conclude that children systematically value dominance. Culture, age and argumentation, can neutralize the inclination towards the dominant individual and even reverse it. For instance, when a dominant's testimony relies on a weak argument or on no argument, preschoolers will rather endorse the testimony of a subordinate based on a strong argument (Castelain et al. 2016). Moreover, the greater allocation of resource to the dominant is also not found for 5-year-old children, and 8-year-olds even favor the subordinate (Charafeddine et al. 2016). Finally, the preschoolers' tendency to see themselves as dominant, and to give greater weight to the testimony of a dominant individual, is not observed in Japan, a culture which values less self-enhancement and which values modesty more (Charafeddine et al. in press).

A limitation of these latter studies, however, is that they do not examine the evaluation of hierarchy with respect to affiliative preferences *per se*. Studies on testimony rather reveal children's epistemic considerations, and allocation tasks rather capture considerations of fairness. Children could however think that the dominant is more knowledgeable or is entitled to more resources, but dislike her/him because of her/his antisocial behavior. Moreover, when children view themselves as more dominant, this does not reveal how they see other's status. In short, the children's valuing a dominant position does not imply that they like dominants more. In this chapter, we thus directly examine children's preferences towards individuals in dominance interactions.

Experiment 1

Method

Participants

One hundred and ten participants (47 four-year-olds and 63 five-year-olds), who were recruited from nursery public schools in Lyon, participated in this experiment. Informed parental consent was obtained for each child. The study received the approval of local board of education (CARDIE: Coordination Académique Recherche-Développement, Innovation- Experimentation). All participants were treated according to the Declaration of Helsinki. Ten participants were excluded from the analysis: 4 children did not want to answer and 6 participants choose both characters. The final set of participants included 100 children (52 girls and 48 boys) from two age groups: 40 four-year-olds (Mean age: 54,2 months; 22 girls and 18 boys) and 60 five-year-olds (Mean age: 66,5 months; 30 girls and 30 boys).

Procedure and Material

Participants were presented with a single story that involved a dominance relationship between two characters who had conflicting goals. The story ended after one character imposed her/his will twice on the other character. In particular, the experimenter said that it was the story about two girls/boys who played together in a park and showed the picture of two girls/boys in a park, on a computer screen (see Figure 13.1). She explained that they had fun together and played for a long time but that, at some point, they wanted to play a different game together and disagreed about which game to play. For instance, the experimenter indicated that one boy wanted to play on the swings and said "Now, let's play on the swings", but also indicated that the other wanted to ride on a toboggan and said "Oh no, now let's ride on the toboggan". Participants were then shown the pair playing on the swings (see Figure 13.1). Later on, they wanted to play another game together but again disagreed (e.g. for instance, playing ball vs. playing with a hoop), and the character who imposed his choice in the first conflicting situation again imposed his choice in this second conflicting situation (e.g. both playmobiles played ball, see Figure 13.1). At the end of the story the two characters appeared without playing and the experimenter asked: "Which girl/boy do you like the most"?

Results

Overall, 51% of the children indicated that they liked the dominant character more and 49% of participants indicated that they liked the subordinate more, a result that does not significantly differ from chance (Binomial test $p = .92$, 95% CI [38.86, 59.19]). The preference for the dominant character did not differ according to the participants' gender (girls: 53.8%, boys: 47.9%, $\chi^2 = .35$, $p = .55$, Cramer's $V = .059$) nor according to the character's gender (female characters: 50%, male characters: 51.9%, $\chi^2 = .037$, $p = .85$, Cramer's $V = .019$). There was also no difference between 4- and 5-year-olds (4-year-olds: 47.5%, 5-year-olds: 53.3%, $\chi^2 = .33$, $p = .57$, Cramer's $V = .057$).

Discussion

In this experiment, participants did not show a preference for one character over the other. The bias towards dominant individuals reported in testimony,

Which boy do you like the most?

Figure 13.1: A sequence of snapshots used in Experiment 1

allocation and self-perception tasks appears to have been neutralized in this explicit preference context, which was designed to better reveal considerations of affiliation. A possible interpretation is that a in such context participants attended more to the antisocial behavior of dominants. Nonetheless, some methodological features may also have limited the attraction for the dominant. First, the social asymmetry might not have been salient enough due to the relative poverty of the interaction between the two characters. Indeed, the two playmobiles were not talking to each other as it was the experimenter who indicated what each playmobile said. The conflict between them and the possibility that one playmobile will prevail, might thus have been attenuated. In addition, it was also unclear how they ended up deciding to play the same game: the dominant did not commend the subordinate to play her/his game and the subordinate did not show any sign of acquiescence. This may not have sufficiently revealed the superiority of the dominant over the subordinate. To overcome these difficulties, Experiment 2 uses a more lively interaction involving two puppets that directly talked to each other to express their conflicting preferences. Moreover, during the exchange the subordinate puppet explicitly indicated that he complied with the dominant's choice, showing thus more clearly the dominant's precedence.

Another issue is that the procedure implemented dominance through one particular form, namely decision power, but other forms of dominance could make the asymmetry more visible through *physical* dyadic interactions. In Experiment 2, we used a play-fighting interaction in addition to decisional power. In this situation, the dominant puppet was the one who imposed his physical superiority over the subordinate puppet. These two inductions of dominance were used in one of our previous studies on dominance perception (Charafeddine et al. 2015). This work showed that 3- to 5-year-old children judged that the puppet that imposed his decision or the puppet that prevailed in the play-fighting interaction was "the boss" ("le chef" in French). We thus also tested 3- to 5-year-old children in Experiment 2. Finally, another change concerned the test question, which more directly involved affiliation concerns.

Experiment 2

Method

Participants

Experiment 2 included 127 children (67 girls and 60 boys) between the ages of 3 and 5, who were recruited from nursery public schools in Lyons. Informed parental consent was obtained for each child. The study received the approval of local board of education (CARDIE: Coordination Académique Recherche-Développement, Innovation- Experimentation). All participants were treated according to the Declaration of Helsinki. A three-year-old girl was excluded from the analysis as she chose both characters. The final set of participants involved 43 three-year-olds (Mean age: 42,7 months; 25 girls and 18 boys), 40 four-year-olds (Mean age: 54,1 months; 20 girls and 20 boys) and 43 five-year-olds (Mean age: 65 months; 21 girls and 22 boys).

Procedure and Material

Each participant was presented with two dominance scenarios based on dyadic interactions that were previously used in Charafeddine et al. (2015). The interactions involved two children embodied by two identical masculine puppets. Two different pairs of puppet were used in the two scenarios. In the decisional power scenario, the two puppets talked to each other and expressed antagonistic desires about games to play together. On two occasions the dominant puppet imposes his will on the other puppet, who then accepted the dominant's choice. Below is dialogue presented to participants (translated from French):

> A (subordinate): Hi Nico!
>
> B (dominant): Hi Léo! What are you playing?
>
> A: I'm playing with marbles. Do you want to play marbles with me?
>
> B: Oh no! I don't want to play marbles. I want so much to play ball. Come, we should play ball.
>
> A: No, but I like to play marbles. Let's play marbles!
>
> B: No, let's play ball, it's so much fun! Come on!
>
> A: Ok! I'm coming!

They were then shown playing with an imaginary ball and the exchange continued as follows:

> B: You see, it's so fun to play ball.
>
> A: Yeah, it's so fun.
>
> B: Ok let's play something else now. Let's jump.
>
> A: No, let's rather run, it's so fun.

B: No, let's jump, it's much more fun. Come on!

A: Ok! I'm coming.

They were then shown jumping together and the exchange continued as follows:

B: See? It's fun to jump!

A: Yeah, it's fun!

In the physical supremacy scenario, two puppets play fight on two occasions and the same puppet prevailed. At the beginning of the interaction the two puppets introduce themselves, indicate that they like to play fight and decide to do so:

A: Hello, my name is Raph. I am very strong and I like so much to play fight.

B: Hello, my name is Thomas. I am also very strong and I, too, like so much to play fight.

Both: Yes! We can play fight together! One, two, three!

Fighting lasted for approximately one minute and ended when one puppet held the other down. After each scenario, participants were asked to choose a child to sit next to him or her. The order of the two scenarios was counterbalanced across participants.

Results

In the decision power scenario, 61.9% of the children chose the dominant puppet (Binomial test, $p = .01$, 95% CI [52.82,70.40]). However, the preference for the dominant puppet differed according to age ($\chi^2_{(2)} = 6.11$, $p = .046$, Cramer's $V = .22$). The partition of the Chi-square showed a significant difference between the 3-year-old group and the two older groups ($\chi^2_{(2)} = 6.09$, $p = .013$, Cramer's $V = .22$) but no significant difference between the two older groups ($\chi^2_{(2)} = .019$, $p = .89$, Cramer's $V = .019$). While among 3-year-olds the preference for the dominant significantly differed from chance (77%, Binomial test, $p = .0006$, 95% CI[61.36, 88.24]), this was not the case for 4- and 5-year-olds: 55% of 4-year-olds preferred the dominant (binomial test, $p = .63$, 95% CI[38.49, 70.74]) and 53.4% of the 5-year-olds did (binomial test, $p = .76$, 95% CI [37.65, 68.82]. The proportions also significantly differed between girls and boys, with boys preferring the dominant puppet more (girls: 53.03%, boys: 71.66%, $\chi^2_{(2)} = 4.63$, $p = .031$, Cramer's $V = .19$). The boys' preference for the dominant puppet significantly differed from chance (71.66% binomial test, $p = .001$, 95% CI [58.55-82.54]) while this was not the case for girls (53.03%, binomial test, $p = .71$, 95% CI [40.34-65.43]).

In the physical supremacy scenario, 61.11% of the children choose the dominant puppet (Binomial test, $p = .015$, 95% CI [52.02, 69.66]). The preference for the dominant did not significantly differ between age groups (3-y.o.: 59.09%; 4-y.o.: 65%, 5-y.o.: 58.13%; $\chi^2_{(2)} = .47$, $p = .78$, Cramer's $V = .061$). None of the results of the three age groups differed from chance (3 y.o.: $p = .29$, 95% CI[43.24,73.66]; 4 y.o.: $p = .08$, 95% CI[48.31, 79.37]; 5 y.o.: $p = .36$, 95% CI[42.21, 72.98]). The preference for the dominant did not significantly differ between girls and boys (girls: 57.57%, boys: 63.33%, $\chi^2_{(1)} = .43$, $p = .51$, Cramer's $V = .059$).

Discussion

For 4- and 5-year-olds, the decision power scenario yielded results that were similar to those obtained in Experiment 1, which involved the same age groups. Indeed, in both experiments, participants' preferences were at chance level. The modifications of Experiment 2, which aimed at making the interaction and the superiority of the dominant more salient than in Experiment 1, did not substantially change the pattern of results for these age groups. However, it appeared that the younger age group—the 3-year-old children—tended to prefer the dominant over the subordinate more than the older age groups. This suggests that participants' evaluation of dominance may change with age. Interestingly, using a resource allocation task and a similar dominance scenario, we previously found a developmental effect that echoes the results found here (Charafeddine et al. 2016). Indeed, while 3- and 4-year-olds allocated more resources to the dominant than to the subordinate, 5-year-olds were at chance and 8-year-olds favored the subordinate (Charafeddine et al. 2016). The greater tendency to prefer the dominant individual in the 3-year-old group could result from a lower ability to empathize with victims and from a greater reliance on authority figures in daily life (e.g. parents, schoolteachers) at younger ages (Charafeddine et al. 2016).

Another interesting result is that the preference for the dominant individual was greater among boys than among girls. Research with adults show that women and men vary regarding their attitude towards hierarchy. Women describe themselves as less competitive than men do and they are also less likely to enter competitive tournaments than men (Niederle & Vesterlund 2007). Moreover, it has been found that men value power more than women while women value status more (Hays 2013). The current result may be due to the early stage of this gender difference in the evaluation of hierarchy. However, both the age and gender differences found in the decision power scenario should be interpreted with caution as the age difference was not replicated in the play-fight scenario and the gender difference was found neither in the play-fight situation nor in the decision power scenario of Experiment 1.

3 Conclusions and perspectives

Taken together the results of these two experiments are mixed and did not reveal a strong preference for the dominant over the subordinate. In the decision power situation (Experiments 1 and 2), four- and five-year-olds' preferences were at chance, though 3-year-olds did prefer the character who imposed his will. Moreover, in the play-fighting situation (Experiment 2), a moderate preference for the dominant individual was found. The preference for dominant individuals, if any, seems thus less pronounced than the preference for familiar, similar, or helpful individuals reported in the literature. The results can suggest that children are weakly sensitive to dominance when manifesting their social preference, but given the ambivalence of the notion, the results can also suggest that children individually differ regarding how they evaluate dominance. Some may be more attentive to the positive aspect, namely the success of the dominant, while others may be more attentive to the negative aspect, namely the costs incurred to the subordinate. To examine further the possibility of such two profiles, future experiments could include more trials per participant so as to examine whether they consistently prefer dominant over subordinate characters across several trials.

A factor that might lead children in one direction or the other and that would be worth investigating is their own dominance rank or how they perceive their dominance rank. If participants occupy a low rank in their peer group, their social experience should lead them to be more aware of the disadvantages of subordination. They could thus experience more negative feelings towards dominants and sympathize more with subordinates. Moreover, given that children are sensitive to similarity, they might prefer individuals whose rank is closer to theirs. In addition to the participants' dominance position, other individual differences might also play a role in children's preference such as their feeling of self-independence, their ability to empathize or their degree of authoritarianism.

Besides personality traits, another aspect that may influence participants' preferences is the context in which affiliations are likely to occur. In the current experiments, participants were asked to make a choice between two individuals, but had nothing to accomplish with those individuals. However, affiliating with a dominant individual might be more efficient in some circumstances than in others. For instance, in a conflict situation or in competitive environments, dominant partners might be helpful allies, while in socially peaceful and stable environments affiliating with them might be less desirable. In conclusion, to better understand preschoolers' social evaluation of dominants and subordinates the consideration of individual factors and the manipulation contextual factors may be helpful.

References

Bar-Haim Y., T. Ziv, D. Lamy & R.M. Hodes (2006). "Nature and nurture in own-race face processing". *Psychological science* 17(2): 159–163.

Bernard S., T. Castelain, H. Mercier, L. Kaufmann, J.-B. Van der Henst & F. Clément (2016). "The boss is always right: Preschoolers endorse the testimony of a dominant over that of a subordinate". *Journal of Experimental Child Psychology* 152: 307–317.

Boehm C. (2009). *Hierarchy in the forest: The evolution of egalitarian behavior.* Cambridge MA: Harvard University Press.

Boyce W.T. (2004). "Social Stratification, Health, and Violence in the Very Young". *Annals of the New York Academy of Sciences* 1036: 47–68.

Brey E. & K. Shutts (2015). "Children use nonverbal cues to make inferences about social power". *Child development* 86(1): 276–286.

Brown D.E. (1991). *Human universals.* New York: McGraw-Hill.

Byrne R.W. & A. Whiten (1988). *Machiavellian intelligence: Social expertise and the evolution of intellect in monkeys, apes, and humans.* New York: Oxford University Press, USA.

Castelain T., S. Bernard, J.-B. Van der Henst & H. Mercier (2016). "The influence of power and reason on young Maya children's endorsement of testimony". *Developmental Science* 19(6): 957–966.

Challman R.C. (1932). "Factors influencing friendships among preschool children". *Child Development* 3(2): 146–158.

Charafeddine R., H. Mercier, F. Clément, L. Kaufmann, A. Berchtold, A. Reboul & J.-B. Van der Henst (2015). "How Preschoolers Use Cues of Dominance to Make Sense of Their Social Environment". *Journal of Cognition and Development* 16(4): 587–607.

Charafeddine R., H. Mercier, F. Clément, L. Kaufmann, A. Reboul & J.-B. Van der Henst (2016). "Children's allocation of resources in social dominance situations". *Developmental Psychology*, 52(11), 1843–1857.

Charafeddine R., T. Yamada, T. Matsui, M. Sudo, P. Germain, S. Bernard, ..., J.-B. Van der Henst (In press). "Cross-cultural Differences in the Valuing of Dominance by Young Children". *Journal of Cognition and Culture.*

Chevaleva-Janovskaja E. (1927). "Les groupements spontanés d'enfants à l'âge préscolaire". *Archives de Psychologie* 20: 219–233.

DeCasper A.J. & W.P. Fifer (1980). "Of human bonding: Newborns prefer their mothers' voices". *Science* 208(4448): 1174–1176.

Edelman M.S. & D.R. Omark (1973). "Dominance hierarchies in young children". *Social Science Information/Information sur les sciences sociales*

12(1): 103–110.
Fawcett C.A. & L. Markson (2010). Similarity predicts liking in 3-year-old children. *Journal of experimental child psychology* 105(4): 345–358.
French D.C. (1984). "Children's knowledge of the social functions of younger, older, and same-age peers". *Child Development* 55(4): 1429–1433.
Gazes R.P., R.R. Hampton & S.F. Lourenco (2017). "Transitive inference of social dominance by human infants". *Developmental science* 20(2): 1–10.
Gülgöz S. & S.A. Gelman (2017). "Who's the boss? Concepts of social power across development". *Child development* 88(3): 946–963.
Hall J.A., E.J. Coats & L.S. LeBeau (2005). "Nonverbal behavior and the vertical dimension of social relations: a meta-analysis". *Psychological Bulletin* 131(6): 898–924.
Hamlin J.K., K. Wynn & P. Bloom (2007). "Social evaluation by preverbal infants". *Nature* 450(7169): 557–559.
— (2010). "Three-month-olds show a negativity bias in their social evaluations", *Developmental Science* 13(6): 923–929.
Keating C.F. & D.L. Bai (1986). "Children's attributions of social dominance from facial cues". *Child Development* 57(5): 1269–1276.
Kelly D.J., P.C. Quinn, A.M. Slater, K. Lee, A. Gibson, M. Smith, L. Ge & O. Pascalis (2005). "Three-month-olds, but not newborns, prefer own-race faces". *Developmental science* 8(6): F31–F36.
Kinzler K.D., E. Dupoux & E.S. Spelke (2007). "The native language of social cognition". *Proceedings of the National Academy of Sciences* 104 (30): 12577–12580.
Kinzler K.D. & Spelke E.S. (2011). "Do infants show social preferences for people differing in race?" *Cognition* 119(1): 1–9.
La Freniere P., F.F. Strayer & R. Gauthier (1984). "The emergence of same-sex affiliative preferences among preschool peers: A developmental/ethological perspective". *Child development* 55(5): 1958–1965.
Mahajan N. & K. Wynn (2012). "Origins of "us" versus "them": Prelinguistic infants prefer similar others". *Cognition* 124(2): 227–233.
Martin C.L. (1989). "Children's use of gender-related information in making social judgments". *Developmental psychology* 25(1): 80.
Martin C.L., R.A. Fabes, S.M.Evans & H. Wyman (1999). "Social cognition on the playground: Children's beliefs about playing with girls versus boys and their relations to sex segregated play". *Journal of Social and Personal Relationships* 16(6): 751–771.
Mascaro O. & G. Csibra (2012). "Representation of stable social dominance

relations by human infants". *Proceedings of the National Academy of Sciences of the United States of America* 109(18): 6862–6867.

Over H. & M. Carpenter (2015). "Children infer affiliative and status relations from watching others imitate". *Developmental science* 18(6): 917–925.

Pun A., S.A. Birch & A.S. Baron (2016). "Infants use relative numerical group size to infer social dominance". *Proceedings of the National Academy of Sciences* 113(9): 2376–2381.

Quinn P.C., J. Yahr, A. Kuhn, M. Slater & O. Pascalis (2002). "Representation of the gender of human faces by infants: A preference for female". *Perception* 31(9): 1109–1121.

Shutts K., M.R. Banaji & E.S. Spelke (2010). "Social categories guide young children's preferences for novel objects." *Developmental science* 13(4): 599–610.

Sluckin A.M. & P.K. Smith (1977). "Two approaches to the concept of dominance in preschool children". *Child Development* 48(3): 917-923.

Thomsen L., W.E. Frankenhuis, M.C. Ingold-Smith & S. Carey (2011). "Big and mighty: preverbal infants mentally represent social dominance". *Science* 331(6016): 477–480.

Chapter 14
How to Be a Direct Realist

Alfredo Paternoster

Abstract My aim in this paper is to argue for a certain version of direct realism, which I take to be the best way of being (simply) realist about the existence of the external world. I start by arguing that realism is best vindicated by direct realism in philosophy of perception. Then I try to show that the best version of direct realism is "relational" realism, according to which real objects are constituents of perceptual experience. Some problems usually ascribed to this view are discussed.

Keywords Direct realism · Representationalism · Tyler Burge

My aim in this paper is twofold. First, I argue that subscribing to direct realism in the philosophy of perception is the best way of being (simply) realist about the existence of the external world. The idea is that, if one wants to be a realist about the existence of the external world (and I take for granted, without justification, that one should want it), one should endorse a direct-realist theory of perception. Second, I argue that providing a convincing justification of direct realism is very hard, since all versions of direct realism (essentially, the representational one and the relational one) face problems. My conclusion will be, therefore, somewhat pessimistic: direct realism is more a requirement for a theory of perception than a properly justified substantive thesis.

In the first section I explain why direct realism is the proper way of vindicating realism. The second section is devoted to a critical discussion of Burge's representational account of direct realism, which I take as an exemplar case study. In the last section I discuss the relational view, arguing that, although it is the best way to account for direct realism, it still faces at least one difficult problem.

1 Setting the stage: (simple) realism and direct realism

Despite a few tenacious opponents, realism about the existence of an external world independent of perceiving subjects (i.e., about the existence of ordinary physical objects occupying places outside our body) is not a position that can be seriously given up. Yet, it is arguably impossible to provide a conclusive refutation of skepticism.[1] However one proposes to cope with skepticism, a crucial role is played by perception: veridicality, or objectivity, of perception seems to be a necessary ingredient of any strategy to vindicate realism. As Bonjour (2007) points out, justifying realism requires both a certain view of the nature of perceptual experience and an account of the relation between experience and perceptual beliefs. To put it roughly, we need a realist theory of perception—i.e., a theory capable of vindicating the ontological subject-independency of what we experience in a perceptual act—and a theory validating the reliability of the process by which perceptual beliefs are derived from perceptual experience. Here, however, I shall be concerned only with the former issue.

Therefore, the question is: what theory of perception can best justify realism?[2] What we want is a theory of perception implying that the ordinary objects that we take as given to us in a perceptual act, such as tables, chairs, books etc., are real (are really out there) and exist independently of us. To be sure, almost all current theories of perception are realist. Indeed, all theories are intended to account for the difference between veridical and non-veridical experience, taking for granted that the veridical case is the ordinary case. Yet, there are different accounts of perceptual experience, resulting in different ways of being committed to realism (or, from a slightly different point of view, in different degrees of commitment to realism). My starting hypothesis is that the most effective way of being a perceptual realist is subscribing to *direct realism*.

Direct realism is the thesis according to which, when a subject has a (gen-

---

[1] I am inclined to think that the best strategy to cope with skepticism consists in showing that it is hard to make sense of it, in a more or less Wittgensteinian vein. See especially his *On Certainty* (Wittgenstein 1969).

[2] Here a *caveat* is in order. One should not interpret "justify" as a sort of demonstration. One cannot provide a theory of perception that *shows* that realism is true, since realism is rather *presupposed* by any realist theory of perception. Yet, different ways of working out the concept of (perceptual) experience result in more or less robust formulations of realism.

uinely)³ perceptual experience, he is in direct contact with objects or layouts of surfaces in the external world. It is very difficult to spell out what 'direct' exactly means, but I think that the idea is intuitively clear: in order to figure out what this directness or immediacy is, you have just to take your perceptual phenomenology at face value. Anyway, I shall specify later on what my requirements for direct realism are.

The reason why I think that direct realism with respect to perception is the best way to account for realism in general is very simple: perception is our primary, fundamental source of knowledge—as Burge effectively puts it, "origins of empirical objectivity lie in perception" (2010: 107). Therefore, the more perception is reliable, the more realism will be warranted, and being committed to direct realism is clearly the most plausible strategy to ensure the reliability of perception. The following remarks on direct realism should provide further evidence for this claim.

There are five *prima facie* reasons to think that direct realism is true:

1. (*phenomenological reason*) Experience presents itself as a direct relation to external objects and properties, in a twofold sense: *a*) what is given in the experience appears to be actual, immediately present and distinct from us (we could call it "principle of actual presence"); and *b*) the experienced properties appear to be properties of external objects (not properties of the experience itself)—this is the well-known "principle of transparency" (see Harman 1990; Martin 2002).

2. (*explanatory reason*) Our action is usually successful. For instance, I can grasp the object I am looking at. The most straightforward explanation of this fact is that we perceive the objects themselves.

3. (*explanatory/evolutionary reason*) The function of perception is probably that of allowing us to access the world in such a way to make our behavior most effective. Arguably, this goal is best attained when what is immediately presented in experience is the world itself.

4. (*epistemological reason*) Direct realism is the account of perception which best grounds the veridicality of perceptual knowledge (this is the reason that led me to introduce direct realism as a position worth defending).

5. (*semantic reason*) Direct realism is the account of perception that best

---

³Hallucinations are not relevant to realism.

justifies our referential uses of words—for instance, the fact that the word 'chair' refers to real chairs.

To be sure, *prima facie* reasons are far from being conclusive reasons. Perhaps the only uncontroversial reason is the first. Indeed it is a platitude, an obvious fact, that in a perceptual act it *seems* to us that we are in an unmediated relation with external objects in the world. But, of course, direct realism is the thesis that in a veridical perceptual event we are *really* in an unmediated relation with external objects. The principles of actual presence and transparency only state that, *ceteris paribus*, we should prefer a direct realist account of experience (with respect to an indirect one), but this can by no means be considered as a mandatory requirement.

Now, what does the expression "to be in an unmediated relation" mean? The most plausible interpretation is the following: what appears to one, or what one seems to perceive, in a perceptual experience (I henceforth assume that 'perceptual' implies a relation with the external world, that is, 'to perceive' is factive)[4] is a real object (or, more precisely, a *part* of a real object—for the sake of simplicity I shall ignore this qualification, though it is by no means a minor point). To say this is not enough, however. In fact, there are at least two families of theories that pretend to be compatible with this claim.

According to what is arguably the most straightforward interpretation, the claim should be intended as implying that the object is a *constituent* of the experience. I shall call this position, which is endorsed by disjunctivists, "object-involving" or "relational" direct realism. Indeed this view is often described in literature as "relationalism". According to the other interpretation, we perceive a real object by *representing* it: on the representational view, in the experience we perceive directly real objects, but these are not constituents of experience.[5]

Unfortunately, the picture is complicated by the fact that relationalism may be compatible with representationalism: if a perceptual state is taken to have a so-called "Russellian" representational content, representational theories turn out to be relational too. In fact, Russellian content includes among its constituents worldly objects and properties.

Now, for the sake of simplicity, I shall not take into consideration the Russellian content view, because, I would say, it compounds the shortcomings of

---

[4] I include in the class of perceptual experiences even most illusions. Therefore, on my view, the great divide is between perceptions and hallucinations.

[5] Searle (2015) suggests that perceptual experiences are *presentations*, rather than representations of the world. Although this correction could seem important (cf. section 2 *infra*), there is no reason to regard Searle's theory as a third position in the field, since it is a standard representational theory under all other aspects.

both the positions. Indeed, as I shall argue in the next sections, I think that direct realism is more "at home" with relationalism, precisely insofar as relationalism (whatever are its problems) is construed as a non-representational account. Therefore, on my view, the Russellian content picture tries to put together two ideas that are better to be regarded as being in opposition. To be sure, the issue would require a much deeper analysis, which I cannot do here.

2 Troubles with representationalism: a case study

In his *Origins of objectivity* (Burge 2010; see also Burge 2005), Tyler Burge has put forward one of the most influential accounts of perception. It is a view in which direct realism, representationalism and empirical results from computational psychology are admirably combined. Yet, as I am going to show, the overall result faces some difficulties. Or, at any rate, direct realism is not easily vindicated.

Let me start with what I take to be the core of Burge's picture: the principle of proximality, which states that:

> Holding constant the antecedent psychological set of the perceiver, a given type of *proximal stimulation* (over the whole body), together with associated internal afferent and efferent input into the perceptual system, will produce a given type of perceptual state, assuming that there is no malfunctioning in the system and no interference with the system. (Burge 2005: 22)

Under these hypotheses, if a change in the distal stimulus (= in the object) is not registered in the proximal stimulus (counterfactually speaking), the experience of the subject will not change.

Burge argues for this claim by discussing the case of two perceptual events that are exactly alike except for the fact that they involve numerically distinct perceived objects. For instance, the event A is the perception of a certain car, while the event B is the perception of a distinct but qualitatively identical car. Since the two cars are type-identical (they share all the properties except their position in the time-space), the two experiential events are also identical. In fact the two objects determine an identical proximal stimulus, thus they cannot be discriminated by the subject.

Therefore, the principle of proximality[6] turns out to be inconsistent with relational direct realism, as it should be expected, since Burge is a represen-

---

[6]As Campbell (2010) points out, perhaps there is a conspicuous idealization in the proximality principle. However, the following "statistical" reformulation could be accepted: in an

tationalist. Indeed Burge takes the truth of the principle as a *reductio* of the disjunctive theory, insofar as science of perception depends on proximality, and so the disjunctive theory turns out to be inconsistent with science.[7] We will go back on this point in the last section, when we will discuss relationalism.

Now, it is well known that Burge is nothing less than the father of externalism in the philosophy of mind (Burge 1979; for the specific case of perception, see Burge 1986). This raises a first problem, because the principle of proximality fits better with the internalist point of view. How can we reconcile Burge's commitment to the principle of proximality with his avowal of externalism?

First of all, let me clarify why I said that the principle of proximality fits better with internalism. The reason is that, according to the principle, any external factor relevant to a perceptual state can be "screened off", so that what determines the content of a perceptual experience is a collection of internal factors after all. In other words, externalism about perceptual content requires that the content systematically co-varies with changes in the environment; but this cannot be the case if the principle of proximality is true, since it is possible that, notwithstanding a difference in external conditions (two distinct objects), perceptual content remains the same (and vice-versa).

Arguably, two replies are available to Burge. Let me first consider a reply that Burge probably would not endorse. The idea is that one thing is the experience and another thing is the content. The "type of perceptual state" mentioned in the principle of proximality is not the content: it is rather the experience itself, or what is usually called the "phenomenal character" of the experience. Therefore content is not affected by the proximality principle, which only concerns the phenomenal character. This view is, however, hardly perspicuous, because it makes perceptual content something detached and independent from phenomenal character. After all, perceptual content is the way the world is given to the subject, so how on earth might it be that external factors are relevant to the content but not to the experience? If perceptual content is the way the world is given to the subject, it cannot be the case that it is not determined by the proximal stimulus. If one wants to take this route, content turns out to be a sort of idle wheel: it is a quite abstract entity whose relation with the notion of what one seems to perceive is far from being clear. On the contrary, in the above-mentioned case of the identical cars, it seems reasonable to say that the content and the phenomenal character are identical too.

---

ordinary context, with nothing of unusual going on, sameness of proximal stimulus in that context is fairly highly correlated with sameness of conscious experience.

[7] Here Burge has in his mind computational psychology.

The second reply is based on Burge's claim that the representational content of perceptual states is partly individuated in terms of what *causes* these states, i.e., external objects. In fact, the existence of the object is required for the existence of perceptual content: the latter depends existentially on its distal cause. This can be *prima facie* interpreted in two ways: Either it means that the representational content can be *described* (and usually is described) in terms of external objects (and their properties), or it means that external objects are *part* of the representational content.

Neither interpretation, however, fits well with Burge's overall theory. The former concedes too much to the objector: in this formulation externalism is not a metaphysical thesis and has no modal import. Externalism turns out to be a very weak thesis, against Burge's view of the matter (see e.g. Burge 1986). The latter amounts to an endorsement of the relational version of representationalism, which he clearly dismisses. What Burge needs is a third interpretation, according to which the existential dependence of perceptual content from the object is a genuinely metaphysical thesis, which, at the same time, does not involve relationalism. It is not clear to me whether there is room for such a position. Be that as it may, it must be conceded that the principle of proximality is more "at home" with internalism.

What are, if any, the consequences of this difficulty for Burge's direct-realist view? First of all, according to Burge (2005: 30), perception is direct insofar as i) the constituents of a perceptual representation *refer* to external items and ii) perception is non-inferential (the transformations operated by, e.g., the visual system, are not inferences). I fully agree on the second point, but the first point is hardly compatible with direct realism. It suggests, in fact, that the perceptual relation between the subject and the reality is mediated by a representation. The idea is that the subject is "in touch" with the object *through the representation*.

Therefore Burge's formulation of direct realism is not entirely perspicuous. It is not clear that he is able to deal with the well known difficulty faced by anyone interested in defending a robust version of direct realism: since in a perceptual act (indeed, in any mental state) the object is always given in a certain way—there is no such thing as perceiving the object "as such"—, it is tempting to say that any perceptual act involves a representation, so that our access to the world is always mediated by a representation. Yet, this is exactly the picture that a genuine direct realist should want to dismantle, because the notion of representation necessarily involves an obtrusive intermediary between the subject and the object. The challenge is to acknowledge the idea that the object is always given in a certain way without succumbing to a form of *indirect* realism.

Faced with this problem, we can be tempted to accept Searle's proposal (see note 5 above) that perception is *presentational*, rather than representational. Actually, this is a merely linguistic amendment: both Burge and Searle take perceptual content as the way the object is given. However, there is in Searle a good suggestion, shared by most representationalists: interpreting the idea that the content of perceptual experience is the way the object is given as a form of indirect realism is the result of a confusion either between perceiving the object and perceiving the experience itself, or between perceiving the object and perceiving the content. Instead, we perceive the object by having an experience—the experience (re)presents the object—, without being in a relation with the experience (unless we reflect on it, but this is not a perceptual state): we are in relation with the object. In other words, to put it in a phrase, we do not perceive the way the object is given to us; rather, we perceive the object in a certain way. And what happens in our head determines the way the object is given, not *what* object is given.

This seems to be enough to save direct realism. Yet, as we are going to see in the following section, there certainly is a more linear way to vindicate it.

3 The relational view (and its shortcomings)

As I have said several times in this article, direct realism requires that what is given in the experience is the real object. According to disjunctivists, only relationalism allows escaping what Zucca (2015) calls the "detachment problem", i.e., the impression that experience is "disconnected" from the world. Indeed, if perceptual experiences are representations, how can we still defend the idea that what is given in the experience is the object? Representations are not world-involving; they are at most world-depending.

The relationalist claim, however, faces immediately a difficulty. How on earth can a real object be part of a mental state? Experience is something internal to the subject. Therefore, whoever wants to be a relationalist must give a non-internal account of experience. He must explain how experience can be outside of the mind.

A good starting point might be subjective evidence: when we have a perceptual experience, something is given to us, but is not given as something mental, or internal (i.e., in the head), whatever this exactly means. It is given as being outside there. And if we try to reflect on our own experience, we are unable to find something "inside us"; we still find the object with its properties, which presents itself as being outside there (cf. the notion of transparency of experience, §1 above). Under this aspect, there is a deep phenomenal difference between propositional thought and perception. And arguably there is also a difference, though less dramatic, between imagery and perception.

Of course, phenomenology is not the last word. It is well known that phenomenology is often deceptive. Yet I think there are certain aspects in phenomenology that should be taken seriously. In particular, I think that if phenomenology deceived us about the issue at stake here, our view of reality would be too shaken. Moreover, even when phenomenology is in certain respects wrong, philosophers usually claim that an account should be given for the phenomenal facts: we can argue that phenomenology is wrong, but we are called on to explain why it goes wrong. If we are unable to do that, we should *prima facie* take phenomenology at face value.

So, we face an apparent difference between the phenomenology of perceptual experience and the phenomenology of thought (including imagery). There seem to be at least two ways of explaining such a difference:

1. Both what is given in perception and what is given in thought are "manufactured" by the brain. "Materials" used in these manufacturings are partly different,[8] and it is this difference that explains phenomenal differences. To put it shortly, there is more information in perception, creating the impression of reality.

2. There is something in perception that is not manufactured by the brain: real objects. The perceptual system was selected so as to (and is organised in such a way as to) keep us in touch, in contact, with the world. The perceptual system simply presents the world, without making use of images or any other kinds of representation. By contrast, when I'm *thinking* of an object, since of course the object is not perceptually available, the thought presents an image or another kind of representation of the object. Even if we (correctly) say that the object of thought is a real object (e.g., when I think of my wallet, it is my wallet "in flesh and blood" that I think of), necessarily, there must be a mental vehicle of the object of my thought. By contrast, perception does not need vehicles (at the personal level).

Both explanations state that there are different processes involved in thought and perception, but they are crucially different under one aspect: Explanation 1 implies indirect realism, since what I perceive is an image created by

---

[8] More rigorously: there are some areas activated during perceptual processing of a certain stimulus (say, a cat) that are not activated when I think of or imagine that cat.

the mind/brain, whereas Explanation 2 involves a commitment to direct realism. In the former, phenomenology is to a certain extent deceptive, insofar as it hides the mental nature (of the contents) of experience. In the latter, phenomenology is not deceptive: what is given to me in perception seems to be out there because it *really is* out there. Note that in the latter case I can say that my thought refers to (or is about) a given object, but I cannot say that my perceptual experience refers to (or is about) that object, because the object is "inside" the experience—the object is constitutive of the experience.

In section 1 I provided some reasons for direct realism. Over and above the already familiar phenomenological and epistemological reasons, I recall the explanatory argument from the success of action: indirect realism is much harder to believe because, if my perceptual contents (understood, neutrally, as what is given in my perceptual experiences) were mental entities, then even my action would be directed to mental entities, and this seems absurd.[9] To put it in a nutshell: *thought is representational, but perception is not*. The role of thought is exactly to "re-create" the world in absence, in order to make plans, figure out how things might be, etc. Representations make possible this goal. By contrast, the role of perception consists basically in making us able to "navigate" successfully in the environment; therefore the world is directly involved. When I am thinking, I can "bracket" the world, but when I am in a perceptual state I cannot bracket the world, it is a metaphysical impossibility.

What I have said so far does not amount to denying that what I perceive depends also on mental operations. As we saw above, the object is always given in a certain way, and the way we perceive it is determined, in part, by the way our perceptual system works. Nevertheless, it is the object that we perceive. The dependence on mental operations does not imply that experiences are representations (more on this later).

Clearly this argument is not enough to persuade representationalists that experience is outside the head, because of the strong intuition that experience—what seems to us to perceive—depends on *internal* operations. And I agree: there is something really puzzling in the conjunctive claim that neural processes underlying experience are internal but experience itself is not.[10] The

---

[9] Actually, the argument requires more elaboration. In fact, according to indirect realism, when I grasp an object, I am "in touch" with the object only through the mediation of a tactile representation. Yet, this seems to me even harder to believe. Note, moreover, that one could construct a semantic version of the argument: if perceptual contents were mental entities, ordinary words such as 'chair' or 'table' would refer to representations.

[10] Clearly, representationalists do not think that direct realism requires paying a so high price. They think that a mental state (in the head) can directly present an external object. We saw in the second section why this is unconvincing to a certain extent: it is an unstable position,

first step to take for not being puzzled is to separate totally the two levels: one thing is what happens in the head when we have a perceptual experience; quite another thing is the experience itself (more on this below).

However, there are still at least two major objections that can be made to relationalism and the related thesis of the external character of experience. The first objection is a variation on the evergreen theme of the argument from hallucination/illusion: there are mental states (e.g., hallucinations) that can be phenomenally identical to perceptual states, and the easiest explanation of this fact is that the two mental states are of one and the same kind. The second objection is that the relational account is in conflict with cognitive science (specifically, with the computational theory of vision).

Although there are familiar answers (set out long since in Austin 1962) to the first objection, I take it as the most harmful and I think that the following, tentative reply is still open to some criticisms. In the first place, the answer to the argument from hallucination is that there is no reason to take the phenomenological indiscriminability as a criterion for identity. A genuine perception and a hallucination are two different kinds of mental state even if they have a common factor. However, this answer does not address the real point: if it is possible that hallucination and perception are *phenomenally* identical, this seems to be a good reason to believe that the phenomenal character of experience, even in the genuinely perceptual case, is fully determined by internal facts—external facts are screened off. And this entails, of course, that relationalism is false: the object plays no direct role in perceptual experience. In other words, the problem is precisely the common factor, the *phenomenological* identity or indiscriminability.[11]

Many words have been spent on this theme,[12] but I am forced to be very brief, so I just give a sketch of what seems to me the picture of the situation. The relationalist could argue that hallucination is a perceptual-like kind of *thought*: its etiology and its causal role are different from perception. Hence, the phenomenal character of hallucination is similar to that of imagery, not to that of perception.[13] Admittedly, however, since the phenomenal identity between a genuine perceptual state and a hallucination is a metaphysical possibility—a sort of stipulation—, it is not something that we can rule out by

---

constantly open to the threat of collapsing on indirect realism.

[11] I will not discuss here the argument to the effect that phenomenological indiscriminability does not imply phenomenological identity, even if my proposal could, in a certain sense, be seen as a variation of this strategy.

[12] See e.g. Martin (2004), Sturgeon (2006), Fish (2009), just to mention a few.

[13] According to Fish (2009) hallucination has not a phenomenal character (so that there is no quest of common factor), but this seems to me too strong.

argument. It must be acknowledged that relationalism has troubles to face this objection.

As to the second objection, I think that it can be presented in two (related) ways. First, the objector complains that the relational account denies the existence of representations, whereas the concept of representation is pivotal in cognitive science. The answer is that the relational account is perfectly compatible with the existence of representations, provided they are conceived of as subpersonal structures, and it is in *this* sense that cognitive science talks about representations. But these "representations" (admittedly, an unhappy expression, though, like almost everybody, I myself have used it many times) are not experiential, that is, are not personal contents, stand-ins for real objects. They are just pieces of information playing a role in certain theories. The point can further be clarified by taking into consideration the other way of couching the objection, the Burgean argument based on the proximality principle. What does it mean that science conforms to the proximality principle? It does not mean that it individuates perceptual states in a non-relational way, since computational vision science is not particularly interested in ordinary (i.e. experiential) perceptual states. There is a sense in which science is committed to a principle of proximality: it is the fact that scientific explanations usually take proximal, rather than distal, causes as prior (this claim should be taken with some caveats that I cannot discuss here). Yet, it seems to me clear that this has nothing to do with direct realism. Direct realism does not concern at all either subpersonal states or scientific explanations.[14]

What I am suggesting is therefore that direct realism is best warranted by a "no-content" view of perception. According to Hutto & Myin (2013), the no-content view implies also the rejection of computational psychology. Therefore they agree with Burge's premise that the relational view is incompatible with computational cognitive science, but draw the opposite conclusion: so much the worse for computational cognitive science. However, their motivations for this conclusion depend clearly on further assumptions (such as the implication from the concept of information to the concept of content) that are far from being uncontroversial. As I have tried to show, endorsing a certain view of perceptual experience (relational rather than non-relational) is independent of the adoption of a certain kind of explanation in cognitive science.

The problem with science, if anything, is that some (most?) scientists seem to endorse *indirect* realism. Take, for instance, the following claims made by the very influential neuroscientist Chris Frith: "my mind can have no knowl-

---

[14]A similar argument based on the (relative) independency of personal states from subpersonal states can be found in Nanay (2015) and McDowell (1994).

edge about the physical world that isn't somehow represented in the brain" (Frith 2007: 23); "even if all our senses are unimpaired and our brain works properly, we have no direct access to the physical world. We may have the sensation of having a direct access, but this is a brain-made illusion" (*ibid.*: 44).[15] Or consider the following quotation from the distinguished Italian psychologist Paola Bressan, who, though quite unwilling to be involved in philosophical puzzles, claims that "The expression *to construct the world* could seem a poetic way of saying, but it is not. When you look around, you don't have the impression of constructing things (...). But this feeling only depends on the great speed and skillfulness of the building process (...). Our experience of objects is entirely created by the brain" (Bressan 2007: 119, our translation from Italian).

Is there a way to reconcile these statements with direct realism? My answer is that there is a tension only if one adopts a brain-centred attitude. The perceptual relation between the subject and the world is one thing; it is quite another what the brain does (computes) in order to sustain this relation. These are two (actually, more than two) different levels of description, and there is no reason to consider the lower level "more real" or "more veridical" than the upper level. True, if one wants to say that only brain facts are real, then direct realism is, like the quoted authors appear to say, an *illusion*. But consider that the brain is at the service of the body (or of the agent), and it is the whole agent in the first place that is involved in the perceptual relation. Direct realism is a thesis concerning the relation between an agent and its environment, not a relation between the brain and its "environment". Therefore the direct character of the relation between an agent and its environment at high (i.e., personal) level can go together the indirect character of the relation between perceptual "representations" and the external world at low (i.e., subpersonal) level.

In order to best understand this view, we should think of brain operations as a machinery allowing subjects to be in touch with objects. In some cases, our action does not even require a rich model of the world—there is no representation of the object, at any level. In other cases (such as categorisation) a rich model is required. But even in this case, what one is in touch with is the real object; the underlying representation determines how the object is seen, not the object itself.

---

[15] In a similar vein, Thomas Metzinger (2003) claims that the content of perceptual experience is so perfect an image of the world that we do not realize that it is an image. The transparency of experience is an illusion ceaselessly created by the brain.

4 Conclusion

I have argued that direct realism is the best way to warrant realism about the existence of the external world and that direct realism is best accounted for by relationalism. However, relationalism faces a number of objections, both naïve and technical. I tried to show how to deal with a couple of these objections, but I think that the reply to the problem raised by the argument from hallucination is not completely satisfactory.

For this reason, it seems to me that direct realism is best characterized as a constraint on theories of perception, rather than as a substantive thesis that can be demonstrated.

I do not pretend either to have refuted representationalism; more modestly, I hope to have convinced the reader that the formulation of direct realism is easier in a relational non-representational account.

References

Austin J.L. (1962). *Sense and Sensibilia*. Oxford: Oxford University Press.
Bonjour L. (2007). "Epistemological Problems of Perception", in E. Zalta (ed.), *Stanford Encyclopedia of Philosophy*, URL=http://plato.stanford.edu/archives/spr2013/entries/perception-episprob/.
Bressan P. (2007). *Il colore della luna*. Roma-Bari: Laterza.
Burge T. (1979). "Individualism and the Mental". *Midwest Studies In Philosophy* 4(1): 73–121.
— (1986). "Individualism and Psychology". *The Philosophical Review* 95(1): 3–45.
— (2005). "Disjunctivism and Perceptual Psychology". *Philosophical Topics* 33(1): 1–78.
— (2010). *Origins of objectivity*. Oxford: Oxford University Press.
Campbell J. (2010). "Demonstrative Reference, the Relational View of Experience and the Proximality Principle". In R. Jeshion (ed). *New Essays on Singular Thought*. Oxford: Oxford University Press. 193–212.
Fish W. (2009). *Perception, Hallucination and Illusion*. Oxford: Oxford University Press.
Frith C. (2007). *Making up the Mind. How the Brain Creates our Mental World*. Oxford: Blackwell.
Harman G. (1990). "The Intrinsic Quality of Experience". *Philosophical Perspectives* 4: 31–52.

Hutto D. & E. Myin (2013). *Radicalizing Enactivism*. Cambridge MA: MIT Press.
Martin M. (2002). "The Transparency of Experience". *Mind & Language* 17(4): 376–425.
— (2004). "The Limits of Self-Awareness". *Philosophical Studies* 120(1-3): 37–89.
McDowell J. (1994). "The content of perceptual experience". *Philosophical Quarterly* 44: 190–205.
Metzinger T. (2003). *Being No-One. The Self-Model Theory of Subjectivity*. Cambridge MA, MIT Press.
Nanay B. (2015). "Perceptual Representation/Perceptual Content". In M. Matthen (ed), *The Oxford Handbook of Philosophy of Perception*. Oxford: Oxford University Press. 154–167.
Searle J. (2015). *Seeing Things as They Are. A Theory of Perception*. Oxford University Press, Oxford.
Sturgeon S. (2006). "Reflective Disjunctivism". *Proceedings of Aristotelian Society* 80 (suppl.): 185–216.
Wittgenstein L. (1969). *On Certainty*. G. E. M. Anscombe & G. H. von Wright (eds). New York: Harper and Row.
Zucca D. (2015). *Defending the Content View of Perceptual Experience*. Cambridge: Cambridge Scholars Press.

Chapter 15
Intentional Imagination and Delusion

Philip Gerrans & Kevin Mulligan

Abstract This paper develops an account of the nature of imagination as a discrete mental process underpinned by a specialised neural and computational architecture. The account integrates evidence from cognitive neuroscience and developmental psychology with philosophical arguments about the nature of imagination. We situate the account against other philosophical accounts and apply it to the understanding of some puzzling phenomena: delusion, pretence and self-deception. We argue that many of the puzzling features of these phenomena arise because they are analysed with a doxastic framework. When the role of imagination in these cases is properly understood these puzzles become more tractable.

Keywords Imagination · Default thinking · Delusion · Pretence · Self-Deception.

> *The highest possible stage in moral culture is when we recognize that we ought to control our thoughts*
> Charles Darwin, *The Descent of Man*

1 Intentionalism and Functionalism

In this paper we provide an account of imagination as a discrete mental process underpinned by a specialised neural and computational architecture (the default network). This specialisation means that, although episodes of imagining share properties of other mental states (the imagistic properties of perceptual states, some of the causal and inferential structure of doxastic states, and the motivational and affective aspects of desire and emotion), imagination cannot be analysed as a species of some other type of mental state. Furthermore, on our account, imagination is not a process best characterised in functional terms. For functionalists a mental state is characterised in terms of the role (actual or counterfactual) it plays, in combination with other mental states, in causing behaviour (Lewis 1972; Block 1981; Churchland 2005). One reason is that our interest here is in cases which imagination, like a cognitive cuckoo, occupies the same functional role as beliefs, making it very difficult to distinguish from the functional perspective. These are the *doxastic borderline* cases which we discuss in the final section.

Consequently our account of imagination is *intentionalist* in the sense described by Crane in his discussion of Brentano's thesis about the intentionality of mental states (Crane 2009). In order to explain the intentional properties of mental states we need to explain their representational properties, or contents, a familiar point in the philosophy of mind and cognitive science. It is another feature of intentional states that the same content can be represented in different *modes*. We can believe, imagine, know or anticipate that p, regret, believe or be ashamed that q for example, where p and q express propositions specifying the representational content of the different attitudes or states. For most functionalists, while representational contents may be intrinsic, modes are individuated in functional terms (Searle 1983).

For intentionalists *both* representational content and mode are intrinsic features of all intentional mental states. Our view exploits the notion of contents and modes *intrinsic* to states of imagination. We do not deny that imagination plays a distinctive causal role and thus can to that extent be the object of a functionalist theory. However we concentrate on the intrinsic properties which enable it to play its causal role.[1]

How then do we distinguish the mind in imaginative mode from the mind in doxastic, conative, emotive or other modes? The essential properties of the imaginative mode are the following:

(i) Imagination is essentially *independent of proximal stimulus*. We do not deny that it can be triggered by sensations (c.f. the cases described by Marcel Proust), memories or beliefs as when daydreams or fantasies are launched by perception. But imagination, unlike perception, does not depend on causation by its intentional object. Triggering of imagination is in fact an instance of essential property (ii).

(ii) Imagination is *associative*, in the classical sense defined by Hume.

(iii) Imagination is *potentially subject to voluntary control*.

(iv) Imagination is *not ultimately responsive the world*.

(v) Imagination *does not come in degrees* (unlike desires, belief, on many accounts, conviction or emotion)

(vi) Imagination is essentially *episodic*. It makes sense to think of both episodic and dispositional, enduring, states of belief but not dispositional, enduring, states of imaginination.

Other mental states and processes share some of these properties (theoretical beliefs are stimulus-independent, memories and emotions are highly associative, hypothetical beliefs generated in calculating alternatives are subject to voluntary control and many desires are unresponsive to facts about the world for example). But *only* imagining and imaginative states have these properties essentially.

We defend this view on three, related, fronts. The first is conceptual. We argue that any state or process with all these properties is an imaginative one. The second strand of the argument is that recent work in cognitive neuroscience discloses a discrete neurocomputational substrate for imagination which confers these properties on episodes of imagining and on states of imagination. The final strand of our argument, which ties the first two together, is that our account of imagination explains features of what we call *doxastic borderline* cases: pretence, self-deception and delusion. These cases

---

[1] Analogously desiring and emoting are distinguished in the first instance by intrinsic modes not functional roles.

are "borderline" because they have some of the functional profile of beliefs: they participate in chains of thought and cause behaviour. As a result, from a functionalist perspective, they are often analysed in doxastic terms as subspecies of belief, partially insulated from holistic processes of belief fixation. The doxastic approach, while it helps explain belief-like features of borderline states generates other problems. How, for example, can a delusional subject "believe" something which so clearly flies in the face of other beliefs and readily available and incontrovertible evidence? How can the self-deceived intend to believe something they know is false? How can children who are able to engage in sophisticated pretence (analysed as "make *believe*") be unable to grasp the concept of belief? These problems are more conceptually and empirically tractable if the behaviour involved in pretence, self-deception and delusion is treated as being based on imaginative rather than doxastic processes. For example our view predicts that episodes of pretence, self-deception and delusion would involve activation in circuitry specialized for imagination rather than rational belief fixation. This prediction is borne out in the case of delusion, the only case for which there is yet a substantial body of empirical evidence.

Thus we are seeking an integrative account of imagination. One which unifies the concept of imagination used by philosophers and psychologists and brings out connections obscured by its use in different theoretical contexts.[2]

1.1 Cognitive Cuckoos and Borderline Phenomena

In defending this view we explain how imaginative states and processes can occupy the functional role of belief, or at least enough of the functional role of belief to cause an interpreter to think that the agent is acting (including thinking, understood as a mental act) on the basis of belief. These are cases in which imagination behaves as a cognitive cuckoo and generates the problem of borderline phenomena. Delusion, self-deception and pretence are the examples we discuss, with most of the focus on delusion. This explanation allows us to situate our view within a recent debate between doxastic (Bayne and Pacherie 2005) and metarepresentational (Currie G. and Ravenscroft 2002; Currie 2004) theorists of delusion. We side with the metarepresentational theorists but argue that they have complicated their account unnecessarily.

Doxastic theorists and metarepresentational theorists share the functionalist idea that when a mental state plays a systemic causal role of a certain sort it must be a belief. Metarepresentational theorists add an extra layer of com-

---

[2]A version of the present account of imagination is given in (Mulligan 1999). For its relation to accounts given by Meinong and Husserl, see Mulligan (forthcoming) and Cairns (1973).

plication. For metacognitive theorists someone who is deluded that p actually imagines p but believes she believes p. They develop this metacognitive account in order to explain how delusions which start episodes of imagination can come to occupy the functional role of belief.

We share with metarepresentational theorists the idea that imaginative states can come to play the causal role of belief: this is on our view what happens in borderline cases such as pretence, delusion and self deception. However we agree with doxastic theorists that we do not need to invoke doxastic ascent to explain how this equivalence or overlap in of functional role arises. Yet we disagree with doxastic theorists that equivalence or similarity of functional role in borderline cases should lead us to treat these cases as instances of belief.

1.2 Incorporation

What then is going on? Our answer is that in borderline cases a thought or non-conceptual episode such as a sensation plays its role in structuring the agent's psychology without ever being subjected to metacognitive evaluation. We call this type of acting *on the basis of* a mental state without metacognitive evaluation *incorporation*.

Incorporation is the mental operation we refer to with the help of the preposition "on the basis of". For example we fix beliefs, emote, form desires and act "on the basis of" perception. In each case mental states and actions combine or form sequences without a rule for their combination or indeed the sequence itself being mentally represented. Similarly we may proceed from thought to thought or thought to action "on the basis of" the initial thought. Often these processes are *consistent* with rules of inference or patterns of belief fixation but it does not follow that those rules themselves play a role in the process by being represented. This can be true for even quite cognitively sophisticated instances of incorporation. Consider the process of conditional judging in which a person *under the supposition that p* ("on the basis of p") judges that q, while not judging that if p then q (judging a conditional content).[3] Similarly, a person who infers from p to q does not *eo ipso* judge that there is some relation between p and q.

To many of a functionalist persuasion such transitions invite a doxastic interpretation: someone who judges conditionally must believe some conditional content. But this collapses a distinction we need to maintain between thinking/acting "on the basis of" p and believing or judging some complex

---
[3] See Edgington (1995).

content. The need for this distinction is shown by the type of borderline case we examine below and the difficulties they produce for doxastic theories.

We claim that in these cases a person *imagines* that p and acts or thinks *on that basis* without it being the case that she thereby believes that p or believes that she believes that p. We draw on and extend Paul Harris' (2000) account of the role of imagination in the life of the child, a paradigm case of acting on the basis of imagination, to the case of borderline phenomena in adult life and give examples of its pervasiveness in pathological and non-pathological conditions.

The idea of incorporation treads a path between doxastic and metacognitive accounts and addresses a worry about the functional role of associative processes for which Tamar Gendler has devised a solution: the concept of Alief (Gendler 2008a; Gendler 2008b). We briefly situate our account in relation to hers.

The rest of the paper has the following structure. Section 2 is an account of imagination which explains its relationship to other mental states, especially doxastic states. Section 3 gives our interpretation of recent work on "default thinking". We argue that default thinking *is* imagination. Section 4 discusses doxastic versus imaginative theories of pretence, self-deception and delusion. We argue that these phenomena arise "on the basis of" imagination, rather than being special cases of belief somehow insulated from rational evaluation. In such cases subjects imagine something and act accordingly. They are in a very similar state to a child who joins in a game of pretend horse riding. Such a child does not believe the broom is a horse, nor does she mistakenly believe that her imagining that the broom is a horse is a belief (Leslie 1994). Rather she incorporates an imaginative state within the game (Walton 1993). She simply imagines the broom is a horse and acts on that basis. This type of incorporation depends on development of a distinctive cognitive capacity for imagination which we now describe in more detail.

2 Imagination and its Cognitive and Non-Cognitive Counterparts

The variety of phenomena described as cases of imagining is enormous – visualising, day-dreaming, thought experiments, supposition, wishful thinking, make-believe fear and admiration, make-believe desiring (Doggett & Egan 2007) Imagining may be visual, auditive, tactile, proprioceptive, affective, conative and also thoroughly non-sensory. It may involve propositional content or it may not (Stevenson 2003).[4]

One reaction to this diversity is to deny that imagination is a unified mental category. One version of this strategy notes that all these different phenomena seem to be versions of other mental states: perhaps daydreams are children of dreams, imaginary percepts really children of percepts, suppositions children of judgements and so on. The difficulty is that in every case imagination differs from its parent state in the respects we identified above: stimulus independence, associativeness, potential for voluntary control, unresponsiveness. For example visual imagining (visualisation) differs from visual perception in being stimulus independent and potentially under voluntary control even where it has not indeed been produced voluntarily. I can intentionally visualise a rabbit in the absence of rabbits but I cannot intentionally *see* a rabbit under those conditions. Indeed even if I am surrounded by rabbits and my eyes are functioning perfectly seeing the rabbits is not directly within my power *in the same sense* in which visualising rabbits is: opening my eyes and looking is within my power in this sense but not seeing. This suggests that imagination is cognitively distinct from perception although it obviously recruits some of the mechanisms which produce perceptual imagery.

Another more promising strategy is to treat imagination as a unified state but explain the diversity in terms of the *objects* of imagination together with the variety of processes involved in representing them. For example on Mike Martin's account of sensory imagining, to imagine seeing a rabbit is to imagine *a seeing* of a rabbitt; to imagine hearing a rabbit is to imagine an episode of auditory perception of a rabbit and so on (Martin 2002).

On this view the objects of sensory imagination are other sensory states and episodes. On our view to imagine a rabbit is to *imagine* a rabbit. One can accomplish this by imagining seeing, hearing, touching, tasting a rabbit and also of course by imagining judging – supposing - that there are rabbits. And so on. But to imagine seeing a rabbit is not to imagine *a* seeing of a rabbit. It is, we might say, to imagine in visual mode. There are as many ways of encountering a rabbit in imagination as there are in real life, but it does not follow, as Martin suggests, that to imagine a rabbit is to imagine another mental state which mediates the encounter.

We do not need to dispute Martin's account here. It is connected to issues in the epistemology of perception and is intended to be only an account of the variety of sensory imagining. But the issue it raises is important. Martin's account recognises the plurality of types of sensory imagination. But it might be generalised. To represent an object in imagination requires us to use the

---

[4] Gendler (2011) provides a comprehensive taxonomy.

mind's representative resources, whether these are concrete, abstract, sensory, imagistic or propositional (Noordhof 2002). These representational resources are shared by different cognitive systems, memory, perception, belief fixation, judgement and imagination. Thus one might think that to imagine is to imagine an F-ing, e.g. a judging, a seeing, a desiring. Imagining is then a type of meta-act. On our alternative view each episode of imagination such as imagining seeing, hearing, judging, admiring, desiring... does indeed have its own *counterpart*—seeing, hearing, judging, admiring, desiring.... But to imagine-F-ing is not to imagine *an* F-ing. Imaginging-F-ing is rather a type of mode.

The counterpart states or processes which individuate imaginative modes differ from imaginative modes in having *congruence conditions*. Congruence conditions comprise satisfaction conditions and/or correctness conditions. Perceptual experience is congruent when it is veridical, belief when it is probable or true. Fear is congruent when the feared object is in fact dangerous. Desires, like all telic attitudes are satisfied when the world matches the content of the desire or when the desires bring about this match (Perhaps desires also have correctness conditions: one's desire to F is correct only if one ought to F given other enabling conditions).

The *intentional structure* of a mental state refers to the way its information processing architecture enables it to meet its congruence condition. For example, visual states have an intentional structure which is a consequence of the way they represent their objects by processing retinal information. Consequently we cannot see colourless objects. We can locate such objects by sound however in virtue of the intentional structure of audition.

Truth conditions are paradigm cases of congruence conditions for doxastic counterparts of imaginings but every *serious* counterpart has its distinctive congruence condition. This turns out to be crucial for imagination. On many, if not most, accounts of imagination it evolved as part of a capacity for planning (Schacter, Addis et al. 2008; Suddendorf, Addis et al. 2009). Being able to imaginatively rehearse actions and consequences before committing to action is a vital cognitive adaptation. Such rehearsals redeploy some of the mechanisms which generate counterpart states. Imagining playing the piano improves performance precisely because it redeploys some of the same mechanisms as actually playing the piano. Imagining being married or changing jobs helps the decision because it rehearses some of the relevant emotional machinery. And indeed as Daniel Gilbert has emphasised many failures of human decision making are the result of limits on imagination (Gilbert 2006). In these cases imagination is limited by constraints on what can be represented by the subject in counterpart mode.

Imagination exploits the intentional structure of serious counterpart states but does not have their congruence conditions. To play its role imagination needs to have some of the intentional structure of its serous counterparts but unlike those counterparts it does not have congruence conditions.

The fact that imagination has serious counterpart states explains the diversity of imagination, the multiplicity of its objects and modes. Its unity is explained by those of its essential properties it does not share with serious counterparts, one of which is that imagination has no congruence conditions.

2.1 Four features of imaginative modes

2.1.1 Independence from a local stimulus

A fundamental feature of imagination is independence of proximal stimuli. Perception requires the presence of the represented object whereas imagination does not. This contrast is not so clear in the case of other types of imagination whose counterparts can also operate in the absence of a represented object. Most obviously we can desire or form beliefs about something which is not present. It is not however part of our claim that stimulus-independence, or lack of physical presence of the imagined object is *sufficient* for imagination. Rather our claim is only that stimulus independence is one of the necessary conditions for imagination.

Stimulus independence leads to the development (in phylogeny and ontogeny) of specialised mechanisms. In order to imagine the mind cannot be occupied by processing incoming perceptual information. Imagination requires the inhibition of the processing of some incoming perceptual information and the selective activation of neural circuitry, with crucial midline hubs, known as the default system. When the default system is active, systems required for perceptual processing of the environmental stimuli are deactivated.

2.1.2 Associativeness

Default thinking, as it is known, has a crucial property distinctive of the imagination. It is associative in the classical sense described by Hume. Episodes of imagination are, in basic cases such as daydreaming, linked by relations of causation, contiguity and resemblance. Once again this is also true of some of the counterparts of imagination. Perception can lead associatively to desire or emotion for example. The associative properties of the default systems explain how imaginative states which do not necessarily depend on a proximal stimulus can nevertheless be *triggered* by perception.

Habitual patterns of association explain how imaginative states are so readily incorporated. Imagining that p fits naturally into the role of believing that p in contexts where there is an habitual pattern of association for p.

This property of associativeness is connected with another feature of imagination we wish to emphasise at this point.

2.1.3 Unruliness

Rational transitions between doxastic states are classically contrasted with associative transitions between memories, thoughts and emotions.

It is worth making an important point about supposition here. Often imagination produces trains of association which mirror those which would be produced by processes of judgement. One can mentally plan a journey producing a sequence of empirically verifiable events, ("After checking in I will have to go through passport control and be molested by a security guard") or follow a logical train of thought (a train of thought ascribed by "p, so q"). This is because habitual patterns of imagination rehearse patterns of thought which may well have been laid down initially by processes of belief fixation leading to judgement (Byrne 2005; Williamson 2005). But when we take the outputs offline they are not subject to the norms of belief fixation which govern serious counterparts. There is an analogy is with perceptual imagination. When I see a striped tiger I must, in optimal conditions, see stripes. As a result, when I visualise one I tend to visualise a striped one since imagination uses information from previous encounters as elements. But I could imagine a spotted one if I chose.

2.1.4 Voluntary Control (and Imaginative Resistance)

We have emphasised the roles of involuntary or habitual trains of association. But we also have control over imaginative associations. We can intervene in a train of thought, imagining a zebra with spots if we wish.

One cannot intend to perceive, believe, desire or emote (other than in the innocuous sense of placing oneself in optimal conditions for inducing the relevant state) but one can directly produce imaginary states provided one has sufficient control of counterpart processes and conditions are right. This idea needs some finessing since theorists have noted that: firstly, some imaginings might not be possible (imagining oneself performing or sympathising with a morally repugnant action for example) as a contingent feature of individual psychology; secondly, some imaginary states appear to arise endogenously.

The first point addresses the phenomenon of *"imaginative resistance"* as it is called (Gendler 2011). Here we concentrate on aspects of the inability to imagine which Gendler calls "won't or can't imagine". In the first case a subject resists an act of imagination which is not inconsistent with the intentional structure of a counterpart state. In the second the resistance follows from the intentional structure of the counterpart and hence is an *architectural* problem as Gendler call it.

Our analysis of the "won't" case involves the notion of hybrid imagination. The architectural cases can be simply explained in terms of in the intentionalist account. In the pure form of imagination we conjoin representations of properties and the only limits on this process are provided by the logical constraint on the way properties fit together. This is *de dicto* imagination. However most imaginative episodes are not pure in this way. When I imagine how my house would look painted a different colour, I visualise the *actual house* painted a different colour. In these cases of what we might call *de re* imagination properties can be rearranged imaginatively but some objects in the episode are represented by us *as actually existing*. These are objects we believe or take to be real, perceive or remember. These hybrid cases include imagination about the self. If I imagine changing jobs I imagine *myself* with some different properties, not an imaginary person.

The phenomenon of imaginative resistance which challenges the idea that imagination is potentially under voluntary control arises in these hybrid cases. Hume described it in the Essay "Of the Standard of Taste" (1757):

> Where *speculative* errors may be found in the polite writings of any age or country, they detract but little from the value of those compositions. There needs to be but a certain turn of thought or imagination to make us enter into all the opinions which then prevailed and relish the sentiments or conclusions derived from them. But *a very violent effort is requisite to change our judgment of manners, and excite sentiments of approbation or blame, love or hatred, different from those to which the mind from long custom has been familiarized* (...). *I* cannot, nor is it proper that *I* should, enter into such sentiments.[5] (Hume 1757/1985: 247, our italics)

Contemporary philosophers point to cases such as the inability to imagine torturing an innocent child in which the inability does not arise from the in-

---

[5]There are, however, voices that see considerable diffcrences between Hume's problem and the "puzzle of imaginative resistance" (this holds even for Walton, who first referred to Hume, but casts doubts about this attribution by now, see Walton (2006).

tentional structure of the episode but seems to stem from facts about the self which cannot be imagined away. These are cases in which as Nichols (2004) puts it " the imagination rebels against certain elements". We can entertain the proposition (*de dicto*) that we could act that way under certain circumstances as an act of supposing, of make believe judging. However when we try and imagine performing the act (an instance of *de se* intentionality) we cannot annex the imaginary property to the self which is doing the imagining.

This phenomenology of imaginative resistance however is only a superficial counterexample to the claim that imagination is under voluntary control. Imaginative resistance follows in the cases considered from the fact that in episodes of hybrid imagination non-imaginary elements constrain the imaginative manipulation of associated properties. In order for me to imagine torturing an innocent child I would also have to imagine away certain aspects of my psychology which would in effect transform the object of imagination into another person. The act of imagination would cease to be hybrid and become an act of pure imagination directed at an imaginary avatar of myself: someone with the same properties *who is not me*. (Nichols gives a different, though not inconsistent account of limits of first person imagination linked to the metaphysics of personal identity. He is addressing the issue of the role of imaginative thought experiment in metaphysical theorising about the self. (Nichols 2008)

Imaginative resistance may also arise from another source: the *intentional structure* of imaginative modes, the limits on what can possibly be represented within that mode given the intentional structure of the serious counterpart. Visualising, for example inherits the constraints to which seeing is subject. One cannot imagine seeing a colourless object. The same is true of other modes.[6] This constraint on imagination is another potential source of imaginative resistance.

2.1.5 Triggering and involuntary imagination

Many imaginary states arise endogenously via an associative process which is not intended. Perception or memory may lead to trains of daydreaming for example. Our point is not that *every* episode of imagination is produced by the will, but that states and episodes of imagination **are** subject to the will in a way

---

[6]One cannot imagine desiring or hoping for a past event. One cannot imagine **seeing a triangle and that its angles add up to** more than 180 degrees. It is often claimed that the correctness or appropriateness conditions of emotions ascribe value-properties, their "formal objects". Thus indignation is only correct if the situation is unjust. If this is plausible, it is not surprising that make believe indignation about a situation, however imaginary, *must* represent the situation as unjust.

which distinguishes them from perception and belief. We are free to imagine the Faculty Board meeting being cancelled as a result of terrorist attack. We cannot, however, judge that it is cancelled unless we have evidence to that effect.

2.2 Alief and Incorporation

The proposal has to deal with an obvious difficulty: how can imaginative states and activities play so many of the functional roles of beliefs without actually being beliefs? The difficulty is especially acute for a functionalist: when causal roles are very similar it does become difficult to separate states identified only by causal role.

Tamar Gendler addresses this problem by introducing a concept of "alief" to describe the triggering of a behaviour by a state which is not doxastically evaluated (Gendler 2008a, 2008b). An animal or child, incapable of rationally evaluating her states, who simply acts on the basis of perception is in a state of "alief" not belief. And as she points out much of adult cognition also consists in this kind of automatic production of behaviour which does not involve the rational evaluation (actual or potential) of the state which initiates it. Furthermore the trigger need not be a veridical perception, it may be an imaginary state. As she says in a discussion of hallucination leading to action:

> What happens in imagination may have (non-pretend) effects beyond imagination — but it does so when the process of imagining activates a subject's innate or habitual propensity to respond to an apparent stimulus in a particular way. (Gendler 2008b: 566)

This is an instance of the process we describe as incorporation of visual imagination. Gendler however treats it as a case of imagination leading to alief, analogous to the way perception can leads to belief. As she says "imagination gives rise to behaviour via alief" (Gendler 2008b: 566).

Our analysis of this case is different. Consider someone who looks into an empty cage and has the experience of seeing a tiger and then flees. A classic hallucination. Intuitively we say that she flees because she takes her experience at face value. This is correct but the question it raises is the nature of this "taking". Doxastic theorists might say that the experience grounds a false belief that a tiger is in the cage for example. Metacognitive doxastic theorists might say that she forms a belief *about her experience* that it is a veridical perception. For Gendler the person *alieves* that there is a tiger in the cage.

We do not wish to argue with Gendler's idea that imagination can lead to behaviour but we note one important point of disagreement: the need to introduce a further cognitive state or process *between* imagination and behaviour in order to explain the intelligence of the behaviour. On our view imagining seeing can lead to behavior in the same way in which serious seeing issues in behaviour. Would Gendler say that we need to interpose an additional doxastic state between perception and behaviour in order to explain behaviour? The more economical and cognitively plausible view of these cases is that behaviour is based on the content of the relevant initiating mental state. In other words that the relevant mental state is incorporated.

Incorporation is all that is needed for most of mental life. Indeed imagination and belief can be jointly incorporated provided they don't produce any drastic disruptions. Perhaps imagining I am a very good tennis player makes me more confident and relaxed when I play, which in turn makes me play better, thus producing evidence for beliefs about my standard of play. When I miss a shot I imagine it was bad luck or the brilliance of my opponent rather than ineptitude, building confidence to attempt the next one successfully. My take on my tennis ability is a fragile package of fantasy and reality: imagination and belief. Only when the coach dispassionately dissects my game do I realise how much of my tennis personality is borderline. Note however that the process of doxastic evaluation need not be all bad. Perhaps I have incorporated negative imaginings into my game which make me reluctant to attempt some shots and these can be replaced by positive beliefs about my ability in those areas. A necessary prelude is imagining myself performing the difficult shots successfully, leading to practicing and eventually believing that I can play them. This might be what coaches and therapists mean when they talk of "fragile" confidence. What they mean is that cognitive control of action is partly imaginary, partly doxastic and that rational evaluation or sudden confrontation with counter evidence disrupts the balance between imagination and belief.

The case is more complicated in the case of *supposition*. In supposition we imagine something contrary to the facts as a way of gathering evidence and exploring alternatives: a form of thought experiment requiring hybrid imagination. Sometimes we even make quasi-judgements in which we evaluate suppositions: following a train of thought to its logical conclusion as in *reductio ad absurdum*. We then use that evidence to form a belief, selecting the best hypothesis. The resultant *judgement* however is subject to doxastic norms. We can imagine its alternative, but unless the evidence changes (including perhaps evidence we entertain in imaginative supposition) we are not free to abandon it.

Similarly we are free to retain suppositions in the case of counter evidence but judgments must be changed. It is in this sense that judgements are involuntary and imagination is not.

To suppose is to imagine judging: without supposition counterfactual reasoning would be impossible. Imagining the double helix was a process of hybrid imagination governed by knowledge about the molecular structure of DNA. Crick and Watson were free to suppose anything about the shape of DNA in the process of generating hypotheses, but only those suppositions consistent with evidence could become judgements. Judgements are essentially responsible to epistemic norms. Pure imaginative supposition is not.

To summarise so far. Imagination is a process with modes whose serious counterparts are familiar psychological processes. The counterpart of visualising is vision, of supposition, judgement. And so on for every episode of imagination. Episodes of imagination are stimulus-independent, not subject to norms of congruence, associative and potentially under volitional control.

Before we turn to the relevant empirical evidence and apply this account there is one more point to make. Imagination and its counterparts interact smoothly in a normal cognitive economy. Most famously, reading and play involve a smooth interplay of imaginative and counterpart processes. Perception can set off a chain of imaginative states. Processes of judgement can involve the imaginative rehearsal of alternative outcomes and hypothesis generation and testing in a process known as supposition.

As the case of supposition leading to judgement shows, a lot mental life consists of episodes of imagination and counterparts which are smoothly integrated and incorporated. The smoothness results from the sharing of some mechanisms between imagination and counterpart states. There will be many cases in which the phenomenology or resultant behaviour produced by imagination or counterpart states is indistinguishable.

Typically the aetiological question does not arise as long as the ambiguity does not cause any problems. Perhaps some people really understand statistical models and have genuine beliefs as a consequence, others only imagine they do. But they agree about climate change. In such cases agreement is not a doxastic phenomenon but a matter of emotional and behavioural consensus, perhaps even of contagious convictions or of contagious make-believe convictions. The same is probably true about most ideological beliefs. Partisans express emotionally scaffolded suppositions, rather than doxastic states, about Islam, climate change, the Pope, third world debt and the bailout of Greece. Their obstinacy and the verbal fluency with which they are articulated, which mimic the functional profile of doxastic states, obscure the fact that the suppositions in question are generated by an associative process which is ultimately insulated from, and not submitted to, the tribunal of rationality and judgement.

3   Imagination and Default Thinking

Recent work in cognitive neuroscience supports the idea that when human attention is not directed on a stimulus in the environment cognition reverts to a default mode: basically daydreaming.

Clearly default thinking is associative: indeed its paradigm case is daydreaming but the circuitry which subtends it (the *default network* as it is known) is also active in dreaming, mental rehearsal involved in planning and personal decision making.

The default network has a crucial property relevant to understanding both imagination and borderline states produced by imagination (Gusnard, Akbudak et al. 2001; Buckner, Andrews-Hanna et al. 2008; Whitfield-Gabrieli, Thermenos et al. 2009). Importantly neurotypical minds show a reduction in default activity (known as task induced deactivation) for both perceptual attention and abstract thinking leading to judgment. Note that these are paradigm cases of non-imaginative or serious cognition *whose states have congruence conditions*. For example in a working memory task a subject watching letters scrolling across a screen has to determine if sequences have been recently repeated. The interval between repetitions and complexity of sequences can be varied to produce different degrees of difficulty. The working memory task is an artificial example of impersonal problem solving. Areas activated (dorsolateral prefrontal) are those required for reasoning and problem-solving tasks which do not require the indexical representation of information (Whitfield-Gabrieli, Thermenos et al. 2009). Solving these tasks requires *decontextualisation* in the terminology of cognitive neuroscience.

In conditions like this, neurotypical subjects deactivate the default network and show increased activation in dorsolateral prefrontal areas. In an imaging study using this working memory paradigm "[i]t appears that the brain alternates between activation of the default network when not engaged in a task and suppression of the default network when engaged in a task" (Whitfield-Gabrieli et. al. 2009: 1280). In the jargon of cognitive neuroscience default networks are *anti-correlated* with networks dedicated to perceptual attention and decontextualised thought.

Default thinking is the screensaver of the prefrontal cortex, constantly turning over autobiographical episodes in a low energy state in between periods of directing attention to a specific cognitive task.

Why not think of default thinking as a form of imagination? Cognitive neuroscientists agree that default thought involves imagination but because the network is also activated by autobiographical memory tasks and other tasks such as planning, which involve autobiographical representation they tend to characterise it as a system specialised for the manipulation of stimulus free episodic representation *in the service of planning and practical* reasoning (Zipf 1949; Schacter, Addis et al. 2007; Botzung, Denkova et al. 2008; Schacter, Addis et al. 2008; Suddendorf, Addis et al. 2009).

For these theorists the ability to detach from the stimulus is a crucial cognitive adaptation which enables humans to plan for the future by imaginatively constructing and rehearsing different scenarios. This adaptation is called Mental Time Travel since it enables humans to preview and review their actions by pre and re-experiencing them, that is, as we used to say, to remember and to expect and predict:

> We are the only animals that can peer deeply into our futures— the only animal that can travel mentally through time, preview a variety of futures, and choose the one that will bring us the greatest pleasure and/or the least pain. This is a remarkable adaptation— which, incidentally, is directly tied to the evolution of the frontal lobes. (Gilbert 2004)

Gilbert's work participates in a reconceptualisation of episodic memory and imagination as part of a unified capacity for cognitive control which supports personal decision-making. Recent research into episodic memory suggests that in fact episodic memory should not be conceived of as the ability to retrieve previous experiences but as an ability to construct experiences by activating perceptual and affective circuitry *in the absence of a stimulus*. The human prefrontal cortex in combination with hippocampal structures allows us to use information about the world gathered in past encounters in order to remember and imagine possible futures, recombining different aspects of experience to produce scenarios relevant to a decision. Thus memory and imagination turn out to be aspects of the same *constructive* process which assembles and reassembles experiences. This is why damage to episodic memory also destroys deliberation and planning (Klein, Loftus et al. 2002; Hassabis 2007). As Schacter et. al. put it in a recent review: " the medial temporal lobe system which has long been considered to be crucial for remembering the past might actually gain adaptive value through its ability to provide details that serve as building blocks of future *event simulations*" (Schacter et. al. 2007: 659, our italics).

These theorists hypothesise that the ability to produce stimulus-independent experiences is an adaptation for planning. However because so much of the information on which they rely is acquired in the course of studies of episodic memory they draw too sharp a conceptual distinction between episodic memory and imagination *despite the fact that they exploit the same mechanisms and produce the same type of experiences.* The theory of mental time travel also depends on the idea that memory is a *constructive* process: the notion of "retrieval" is a metaphor because the representations which provide memory with its content need to be constructed anew on each occasion. The consolidation of memory is just the overlearning of this constructive process.

Thus what these theorists call episodic memory is really hybrid imagination, the objects of which are past events and states of affairs. Just as we try to work out what *might* happen by imagining the future within contextual constraints supplied by propositional knowledge we can try and work out what *actually* happened to us in the past in exactly the same way. We prompt our imagination and it yields an experience. Often what we imagine happened actually happened, in which case we call it a veridical memory. Often we imagine that something happened to us when it did not. We call this a non-veridical memory but a more accurate description is that we imagined that something happened to us when it did not. If this seems a little counterintuitive consider the parallel account of hallucination. If we look into an empty cage and hallucinate a rabbit we will say intuitively we "saw" a rabbit. In fact we imagined a rabbit by visualising it. Similarly when we try and recall where we were on a particular occasion we imagine ourselves present at that occasion, and say we "remember" being there.[7]

Shacter et. al. (2007-2008) situate their account of mental time travel as part of an explanation of planning involving imaginative rehearsal. If in fact stimulus free thought evolved as an adaptation which enables planning then the crucial adapation for humans is the ability to shift from undirected default thinking to goal-directed imagining of past and future selves and situations. When there is no such goal to which imagination needs to be directed, or no urgent environmental or abstract problem which requires inhibition of default circuitry, the mind reverts to default mode in which imagination continues undirected: screensaver mode.

---

[7]This account of episodic memory is not necessary for our argument but it is highly congenial. We note that it was anticipated nearly a century ago. In his 1900 *Vorlesungen über Psychpathologie in ihrer Bedeutung für die normale Psychologie*, (Lectures on Psychopathology in its significance for normal psychology) Gustav Störring refers to cases of "memory-illusion in which free phantasy presentations which have no basis in past experience are taken as memories" (Störring 1900: 2689).

What cognitive neuroscience calls default thinking underpinned by specialised neural circuitry meets the criteria for imagination, pure and hybrid. It involves the deployment of counterpart mechanisms in many modes to produce states which are stimulus independent, potentially voluntary, associative and not essentially governed by norms of epistemic congruence.

The products of default thinking can become beliefs. Consider a parent whose teenage child has not returned home after a party. The thought occurs "she has had an accident". This thought is a default thought, an imaginative supposition triggered by noticing that it is 4am, rather than a serious, evidence-based judgment about the probabilities. There are different ways such a thought can be incorporated depending on the mode in which the mind processes it. In doxastic mode it is treated as having congruence conditions and potentially an object of judgement. In this mode the mind treats the thought as an abductive hypothesis which, if it were true, would explain the situation. The mind then needs to evaluate it accordingly. If it is judged to be congruent (a reasonable hypothesis according to prior beliefs and the situation) then that supposition generated by imagination can become a belief. Such beliefs then ground enduring multitrack dispositions to behave, to infer, to emote, based on the content of the belief. And in fact this sequence is a fairly natural way to reconstruct belief fixation. Bayne and Pacherie (2005) make a similar point in their discussion of what they call indicative imagination:

> Indicative imagination, by contrast, is routinely triggered by perception. One looks at a colleague's face and imagines that they are anxious... [such] indicative imagination is continuous with belief. To imagine P in the indicative sense is to think that P might actually be true, it is to take P as a contender for the truth. Someone who leaps to conclusions—someone who fails to evaluate a hypothesis in a careful and considered manner—might easily proceed from regarding P as a contender for the truth to thinking that P is true. (Bayne and Pacherie 2005: 170)

However it is not obligatory to treat indicative imagination this way. Someone who sees *p* and thinks *q on that basis* need not do so in virtue of thinking that *q* might be true (although they *could* of course take that attitude to the content).

To return to the case of the anxious parent, worrying about the child, self recrimination, telephoning the child's friends can all occur "on the basis of" the initial thought. Such associative processes do not necessarily require the adoption of the thought as a candidate belief. This type of incorporation represents an attempt to palliate imaginatively-generated anxiety rather than fix a belief. And in fact we often try to reassure such people by reminding them that their worries are generated by imagination and there is no need to believe what one imagines.

If we conjoin the three claims:

(i) default thinking is ubiquitous

(ii) incorporation of default thinking is ubiquitous

(iii) default thinking is imagination

it follows that there is a lot more imagination involved in human cognitive life than people may care to acknowledge. We see no difficulty here and in fact a rather elegant explanation of the pervasive prevalence of irrationality in human life. If people act on the basis of imaginative states, i.e. states produced by an associative process not essentially subject to epistemic norms it is not surprising that epistemic norms do not apply straightforwardly to their behaviour.

4 Doxastic Borderline Phenomena

4.1 Pretence

Childhood pretence has long puzzled developmental researchers. On the one hand it is extremely sophisticated. The child who joins in a game pretending the broomstick is a horse is not taking perception at face value, but imagining something contrary to fact. Children aged 2 can do this and also engage in role play in which they pretend to be someone else, taking up an alternative character and perspective on the world.

Not only that but the child is able to deploy these decoupled representations in combination with other beliefs and perceptions to play the game, updating the pretence appropriately and situating it in a context partly determined by factual beliefs about the world. Children are very good at including in the pretence only those representations which are consistent with the initial stipulations and disallowing others. The horse can't drink sand but it can drink imaginary water, the bad monster behaves in character and so on. These are very sophisticated cognitive abilities probably unique to humans.

Consequently for some theorists pretence involves belief-like states *treated as true within the context of the game*. On this view Pretence represents a stage in conceptual development prior to acquisition of the concept of belief: a type of quasi belief (Leslie 1987; Leslie 1994).

A difficulty for the doxastic view of pretence is that sophisticated pretence arrives so far in advance of conceptual understanding of belief. For example before ability to pass the false belief test and also, interestingly, before the ability to deceive, understood as the ability to induce a false belief in the mind of the other person. Three year olds who can pretend cannot yet lie. And though they can mislead (for example by hiding something) three years olds are not yet masters of deception in the full blown sense of intentionally producing false beliefs in the minds of the deceived. Rather they manipulate their actions.

Like Paul Harris (2000) we take the arrival of pretence to indicate the coming on line of a capacity for *imagination*. In pretence the child imagines something counter to reality and then continues to *act* accordingly. Her default attitude is that pretence is like reality except in those aspects stipulated by the pretence.

We adopt from Harris the point that the child distinguishes pretence from reality, not by mastering a theory about the difference between pretence and belief states which involves metarepresenting the first order states but from the context. The child is able to imagine, to believe and to deploy these states appropriately well before she is able to form judgements about whether or not a state is imaginative or doxastic and whether norms of veridicality or truthfulness apply.

> Young children can engage in pretence and even join in with another's pretence, even if they do not possess any insight into the mental state of pretending. In much the same way, they can hold a belief even if they lack insight into the mental state of believing... Pretence is *not* an early manifestation of the child's developing understanding of mind. (Harris 2000: 192)

On Harris' view pretence is acting on the basis of imagination. A paradigm case of incorporation as we have described it. The ability to distinguish imagination from belief by taking a meta attitude towards the respective states arrives later as part of a package of related capacities which depend on metarepresentational and theoretical capacities which are slower to mature.

The case of childhood pretence shows that factual beliefs can provide raw material for imaginative supposition in the minds of young children without the two states being confused.

The same is true of much of adult life. Books, films and religious rituals depend on the ability to pretend, incorporating pretend mental states while patrolling the boundary between pretence and belief. William Dalrymple's *Nine*

Lives gives many examples of this. The same man who by day is a prison guard or a farmer can become a shaman or an incarnate God during a ceremonial performance. Worshippers know that he has four children and two wives in the village, and that his song protests caste injustice using specific information acquired in his daily life. But he is also a God, *during the performance*. Belief and pretence coexist smoothly precisely because context stipulates the boundaries of pretence. Just as it makes no sense to ask the child " riding" a broomstick " how can a horse be made of wood? " it makes no sense to ask "how does Krishna work at the prison during the week?"

Thus imagination and belief fixation *coexist* in both maturing and mature minds because they are distinct cognitive capacities. Although distinct they are complementary. Once a mind is mature it can use doxastic imagination to provide raw material for judgement. However the use of such supposition *as the basis for judgement* cannot occur until the child can make judgements. That is, choose between suppositions on the basis of epistemic principles.

Cognitive maturation is a process which is not complete until late adolescence. Its crucial morphological feature is the myelination of prefrontal *dorsolateral* areas and the ability to activate those areas in conditions which require the manipulation of context independent representations, that is, in other words, to control the anticorrelation between dorsolateral areas required for decontextualised thought and ventromedial areas which underpin imagination and self-directed rumination. Until the dorsolateral areas develop such anticorrelations are impossible, and children remain prisoners of default thinking. Even once they are developed it takes considerable inhibitory resources to shut default thinking down and shift to a decontextualised mode of thought implemented by dorsolateral prefrontal areas.

This anticorrelation is in fact difficult for most people to accomplish. Decades of psychological research have shown that decontextualisation is a precarious cognitive achievement. Most people make only a partial transition from default thinking when confronted by a problem. Typically they generate imaginative suppositions prompted by the problem. However it seems that the ability to take the next step and evaluate competing suppositions in a decontextualised way requires a high degree of cognitive control which is not always present. Thus there are many cases in which people simply incorporate a supposition. And what they think of as reasoning is simply talking while oscillating between suppositions until one or other is incorporated on the basis of causes which remain introspectively opaque.

4.2 Delusion and Imagination

Delusions provide a nice example of inability to decontextualise, or to incorporate decontextualised conclusions. Delusions are standardly analysed doxastically: as false beliefs produced by abnormal reasoning. They are maintained in the face of obvious evidence to the contrary and often coexist with an intact capacity for procedural reasoning on non-delusional topics. Hence they are not simply corrigible mistakes and in fact the attempt to correct them can lead to considerable resistance, distress and even the consolidation of the delusion. People with paranoid delusions for example may start to treat their doctors as part of the conspiracy against them. Delusions are psychiatric disorders precisely because they exhibit what we called *borderline* properties. They appear belief-like in many respects, prompting trains of thought and behaviour but in other ways, especially their intractability to evidence and argument, do not behave as beliefs are supposed to.

There are now many doxastic theories of delusion which attempt to explain the nature of this coexistence (Maher 1992; Garety and Hemsley 1994; Young 1999; Bermúdez 2000; Gerrans 2001; Broome 2004). The most promising of these identify the neural substrates of cognitive processes which could affect rational belief fixation, insulating a delusional subset of beliefs from rational revision. Thus for example some theorists have proposed that right hemisphere malfunction leads to a failure of "belief evaluation" (Coltheart 2007) and others that dopaminergic dysregulation makes some experiences highly salient, effectively exerting a confirmation bias on belief fixation (Smith et al. 2006). These theories are not incompatible and are in fact complementary. Highly salient experiences could prompt hypotheses which are then confirmed by biased reasoning (Mujica-Parodi et al. 2000).

Other approaches[8] have suggested that delusions arise not through a *failure of reasoning* but through *failure to reason*. Work on the default system in schizophrenia supports this conclusion. In the same study in which Whitfield Gabrielli et. al. established that neurotypical subjects inhibit the default network in order to perform an abstract working memory task they compared the performance of neurotypical subjects, patients with with early onset schizophrenia and first degree relatives. They found "hyperactivity and hyperconnectivity of the default network" in the at rest condition of schizophren-

---

[8]Investigators like Vinod Goel, after reviewing the neuropsychological literature on reasoning in non deluded and deluded (schizophrenic) subjects suggest that the problem for schizophrenic subjects is not reasoning *per se*. Rather as Goel puts it they have difficulty "*accessing the belief-laden reasoning system* associated with the frontal temporal lobe system" (Goel 2004: 88).

ics compared to normal subjects. They also found that the schizophrenic and first degree relatives showed reduced deactivation of the default network in the working memory task and reduced anticorrelations between medial prefrontal cortex (part of the default network) and dorsolateral prefrontal cortex (activated for the impersonal working memory task).

The working memory task is not a reasoning task *per se*. But decontextualised working memory is a necessary condition for procedural reasoning. Subjects who need to reason about the truth need to inhibit incorporation of default personal and emotional associations and evaluate them for epistemic congruence. The natural interpretation of these results is that failure to inhibit the default network produces a failure of decontextualisation necessary to form judgements.

Doxastic theorists of delusion could agree that delusions are initially generated by imagination. A delusion, like many hypotheses, might start life as a doxastic supposition. Often the imaginative process is triggered automatically. Currie and co give the example of seeing an acquaintance who has lost a lot of weight and thinking "he has cancer".

On the doxastic interpretation someone with a delusion then adopts this delusional hypothesis as a belief. But the price for this interpretation is the need to explain how such a belief, if it is a belief, can be insulated from rational evaluation. The natural reply is that the transition from hypothesis to judgement is somehow distorted, by the salience of the hypothesis, its emotional associations, pre-existing biases or a combination of factors. And indeed from within the doxastic framework this interpretation is reasonable.

However from the point of view of cognitive neuroscience it appears that distorting influences do not operate on processes of judgement but on imaginative processes. The "hyperactivity and hyperconnectivity" of the default network generates the delusional thought and prevents it being from referred to the tribunal of procedural rationality. If it is the case that judgement requires inhibition of the default network which produces associative imaginings then judgement cannot arise in cases where that network cannot be inhibited. And if this is the case it is not surprising that delusions coexist with a capacity for judgement in other cases.

The imaginative theorist accepts this account of the genesis of delusion: it is an imaginative response to experience. But it offers a different account of the transition to delusion: the delusion is not submitted to a process of hypothesis confirmation but is simply incorporated.

Note that the process of incorporation need not be a conscious explicit one: very few such acts of incorporation are.

4.2.1 Currie & Co and the process of incorporation

A problem for imaginative accounts is to explain how delusion can continue to play such a belief-like role, structuring behaviour and psychology unless it actually is a belief. It is to answer this challenge that Currie and co develop their metadoxastic account in which a delusion is an imaginary state which the subject *mistakenly believes to be a belief.*

But this suggests that in order to incorporate a state a subject must take a doxastic meta-attitude to it. Cleary there are occasions when this will be the case. Someone with an hallucination who imagines something that is not the case will eventually be confronted by the fact she is acting on the basis of imagination not perception and perhaps will come to believe of her imaginary state that it is in fact an imaginary state. However it does not follow, even in a case like this, that the true metabelief corrects a false first order belief. She does not come to believe of a perceptual belief that it was false. She simply comes to believe of an imaginary episode that it was in fact an imaginary episode.

Currie & Co suggest however that something like the reverse happens in the case of delusion. The subject forms a belief about her imaginary state, namely that it is a belief—perhaps because, as they put it, beliefs are distinguished from imaginings by being action-guiding and not subject to the will. When the subject introspects she encounters a state which is action-guiding and feels involuntary. So she *concludes* it is a belief.

It seems more likely that in incorporating imagination the subject is simply *immersed* in it. Harris gives a very instructive account of children's fear of imaginary monsters and conversation with imaginary friends. He points out that the fear does not abate when the parent puts on the light and shows the child that the room is empty or suggests that the child is imagining it. Nor do children stop conversing happily with their friend. These three-year-olds apply a distinction between imagination and belief and make it for other objects in other contexts but there are some contexts in which they simply incorporate their imaginings. The point is just that in so doing they do not mistake the nature of their mental state. They simply incorporate it without making it the object of any further mental state and so without evaluating it.

A point made by theorists like Bayne and Pacherie against the imaginative account of Currie and Co is that the states in question, unlike imaginary states are not voluntary. And as we noted before it is a hallmark of beliefs that they are not under voluntary control.

But what is meant by this criterion is not that all imaginary or borderline states are *deliberately* imagined and maintained. If this were the case daydreaming would not be an act of imagination. Rather we mean that we can distinguish imagination from judgment in terms of freedom to imagine an alternative. Fantasy and reverie can arise from an endogenous associative process: however we can also turn daydreams off so to speak and imagine the opposite. We do not have this luxury with beliefs formed by judgemental processes.

The apparently involuntary aspect of delusion thus should not make us equate it with belief. This would be the equivalent of suggesting that someone with a persistent hallucination (which originates endogenously) "mis-sees". Or on a metacognitive account that such a person believes that they see. While this kind of mistake might arise if the person was forced to make a judgement about their experience there is no need to suggest that in initially incorporating an hallucination the subject *judges that they see*.

Similarly with doxastic imagining. In order to incorporate it the subject need not take any attitude to it at all. *Forced* to take a meta-attitude the subject might make a mistake and misidentify it as a belief but it is striking how few clinical accounts suggest that this captures anything like the phenomenology or aetiology.

Instead most accounts focus on the way the delusional subject engages with her delusion, becoming immersed in it and the associations it prompts. At the same time she is often able to make judgements on unrelated topics.

The parallel with the child afraid of monsters is relevant again. Such imaginings are endogenous but not involuntary (she can imagine the opposite). Their incorporation does not involve an inability to distinguish imagination and belief.

The deluded adult is not a child but the point of the example is merely to show that doxastic and imaginative states can coexist and interact in even very young minds. This type of coexistence and interaction sustained into adulthood underwrites much of human social life and communication. Story telling, fiction and ritual are ancient human capacities which depend on this kind of incorporation. None of these essentially social activities involve mistaking imagination for belief.

To summarise: Delusions arise via a process of imagination which produces the imaginary state which is subsequently incorporated.

4.3 Self Deception

It is no coincidence that people who have written on imagination and pretence have drawn connections which the phenomenon of self-deception: another borderline case in which doxastic analyses run into problems because of a tension between functional role and rationality constraint. The self-deceived person behaves *as if* she has a belief which, *if she were rational*, she could not intend to form or maintain.

We cannot do justice to the many theories of self-deception here but we do note that the structure of our account devised for other borderline cases extends to self-deception. Here we sketch the outline of a prolegomenon to a future imaginative theory of self-deception.

All accounts of self-deception involve some mechanism which partitions the belief system in order to explain the presence of strongly held, emotionally valent, but inconsistent beliefs. The incompetent administrator who lowered the entry standards to university blames the teaching staff for the high attrition rate in first year medical science. The ego-dystonic belief that one is responsible sustains the defensive belief that others are responsible.

Strong Intentionalist accounts model the partitioning as a deceptive relationship between two people, with one misleading the other. No matter how this account is weakened it seems to involve both an intention to believe that *p*, which causes the belief that *p and* a belief that not *p*, within the same psychology.

Deflationary accounts avoid this problem by treating self-deception as a special case of bias in evidence gathering and assessment rather than a consequence result of beliefs about the self, insulated from rational revision. Opponents of the deflationary view do not dispute a role for biasing mechanisms but they note that the deflationary view seems to miss the central role of the self in producing self-deception.

On our view the self-deceived administrator repeatedly imagines in a variety of ways that he is competent and incorporates this imagining into his behaviour. There is no paradoxical aspect in imagining favourable things about oneself which are inconsistent with the evidence. Imagination is not doxastically congruent. Nor is there any psychological implausibility in the suggestion that imagination is influenced by motivational states produced by emotions. It is a nice corollary of this view that one could perhaps explain his inability to process evidence of his own incompetence doxastically in terms of *imaginative resistance*. He can't generate the supposition (*de se*) that he is incompetent and hence the hypothesis is not really a live one for belief fixation.

Doxastic theorists of self-deception would perhaps agree that the problematic belief originates as an imaginative supposition generated in self defence. However given that the supposition ultimately issues in behaviour they treat it as a belief generated by processes of belief fixation. The problem then becomes explaining the cognitive insulation of the belief.

The problem for the imaginative theorist is different. There is no problem explaining the insulation of imaginative supposition from belief fixation: they have different intentional structures. Here we agree with Nichols and Stich (2006) that the problem of quarantining as they call it is solved by the architecture of the imagination. The problem for the imaginative theorist is explaining how imagination generates behaviour. This problem however is only imaginary. The incorporation of imagination into the psychological and behavioural economy is a fact of cognitive life from toddlerhood to senility.

5 Conclusion

We have argued that imagination is a distinctive cognitive process, with its own architecture, which exploits other cognitive systems to generate representations with distinctive intentional properties. The combination of content and mode intrinsic to imaginative states distinguishes them from other mental states. It is important to our account that the imaginative mode is distinguishable from other modes (doxastic, emotional or conative) not just in terms of its functional role. Of course in general, imagining that p plays a different causal role than believing that p. However, on our view, the two states can be distinguished in virtue of their intentional structure. In particular the lack of congruence conditions for episodes of imagination distinguishes them from their serious counterparts whose intentional structures derive from their congruence conditions. This is not an accidental feature of imagination: specialised neural circuitry has evolved to enable humans to construct and manipulate representations which have representational contents but no congruence conditions, conferring distinct advantages on the human species.

Although they lack congruence conditions, imaginative episodes can be *incorporated* into an agent's psychology, structuring behaviour and patterns of thought. In these cases imagination behaves in a very similar way to belief. In everyday speech we say that people act "as if" they believed a proposition to explain many types of behaviour (play, pretence, engagement with fiction or film) and much philosophical theorising about imagination involves explicating this as if attitude to propositions. Some of this theorising has been extended to explain behaviour in other areas, particularly psychopathology. Such extensions finesse an intuition about, for example delusions, that they are in fact an instance of mistaking an episode of imagination for a genuine perception or belief.

The notion of incorporation subsumes this approach while avoiding some of the complications it produces. Incorporation turns out to be a very fundamental mental operation in which psychology is structured "on the basis" of a mental representation without it being the case that a subject takes an attitude, or even has the disposition to have an attitude towards, that representation. It the concept of incorporation is incorporated into our explanatory repertoire it becomes easy to see how imagination can structure an agent's psychology without it being the case that the agent believes the imagined content or makes a mistake about the process which produces it.

An important aspect of the incorporation account is that it does not require that a subject explicitly know she is incorporating an imaginary episode. Many acts of incorporation are involuntary as anyone who has cried in a movie could attest. The boundaries between incorporated beliefs and imaginings are not always clear to introspection, especially since imaginings can lead to beliefs (when suppositions lead to judgments) and beliefs to imaginings. As the example of the tennis player showed, prising the phenomena apart can be difficult.

We argued that pretence, delusion and self-deception are all cases in which people incorporate episodes of imagination. We do not of course wish to claim that *all* such cases are incorporated imaginings. We accept that people can be acting on the basis of obstinately entrenched false beliefs about themselves and the world. We also accept that at face value it will be very difficult to distinguish doxastic from imaginative incorporation. Nonetheless we need to make conceptual space for those cases in which people are hostages not to biases in belief fixation but to their own entrenched tendencies to imagine.

References

Bayne T. & E. Pacherie. (2005). "In Defence of the Doxastic Conception of Delusions Mind and Language." *Mind and Language* 20: 163–188.

Bermúdez J.L. (2000). "Normativity and rationality in delusional psychiatric disorders." *Mind and Language* 16(5): 457-493.

Block N. (1981). "Psychologism and Behaviorism." *Philosophical Review* 90: 5–43.

Botzung A., E. Denkova et al. (2008). "Experiencing past and future personal events: Functional neuroimaging evidence on the neural bases of mental time travel." *Brain and Cognition* 66(2): 202-212.

Broome M. (2004). "The Rationality of Psychosis and Understanding the Deluded" *Philosophy, Psychiatry, & Psychology* 11: 35–41.

Buckner R. L., J. R. Andrews-Hanna et al. (2008). "The brain's default system: Anatomy, function, and relevance to disease." *The Year in Cognitive Neuroscience 2008, Ann. NY Acad. Sci* 1124: 1–38.

Byrne R.M.J. (2005). *The rational imagination: How people create alternatives to reality*, Cambridge MA, The MIT Press.

Cairns D. (1973). "Perceiving, Remembering, Image-Awareness, Feigning Awareness." In F. Kersen & R. Zaner (eds) *Phenomenology: Continuation and Criticism, Essays in Memory of Dorion Cairns*. The Hague, Netherlands: Martinus Nijhoff. 251–262.

Churchland P.M. (2005). "Functionalism at forty: A critical retrospective." *The Journal of philosophy* 102(1): 33–50.

Coltheart M. (2007). "The 33rd Sir Frederick Bartlett Lecture Cognitive neuropsychiatry and delusional belief." *The Quarterly Journal of Experimental Psychology* 60(8): 1041–1062.

Crane T. (2009). "Intentionalism." In A. Beckermann and B. McLaughlin (eds). *The Oxford Handbook to the Philosophy of Mind*. Oxford: Oxford University Press: 474–93.

Currie G. & I. Ravenscroft (2002). *Recreative Minds*. New York: Oxford University Press.

Currie G. & J. Jureidini (2004). "Narrative and Coherence." *Mind and Language* 19: 409–427.

Doggett T. & A. Egan (2007). "Wanting Things You Don't Want: The Case for an Imaginative Analogue of Desire". *Philosophers' Imprint* 7(9): 1–17.

Edgington D. (1995). "On Conditionals". *Mind* 104(414): 235–329.

Garety P.A. and D.R. Hemsley (1994). *Delusions: Investigations into the Psychology of Delusional Reasoning*. Oxford: Oxford University Press.

Gendler T.S. (2008a). "Alief and belief." *Journal of Philosophy* 105(10): 634–663.

— (2008b). "Alief in action (and reaction)." *Mind & Language* 23(5): 552–585.

— (2011). "Imagination", *The Stanford Encyclopedia of Philosophy (Fall 2011 Edition)*, Edward N. Zalta (ed). URL=http://plato.stanford.edu/archives/fall2011/entries/imagination/.

Gerrans P. (2001). "Delusions as performance failures." *Cognitive Neuropsychiatry* 6(3): 161–173.

Gilbert D.T. (2004). "Affective forecasting ... or ... the big wombassa: what you think you're going to getm and what you don't get, when you get what you want", *Edge*. URL=https://www.edge.org/conversation/daniel_gilbert-affective-forecastingorthe-big-wombassa-what-you-think-youre-going-to

— (2006). *Stumbling on happiness*. New York: Knopf.

Gusnard D., E. Akbudak et al. (2001). "Medial prefrontal cortex and self-referential mental activity: relation to a default mode of brain function." *Proceedings of the National Academy of Sciences* 98(7): 4259.

Harris P. (2000). *The Work of the Imagination*. Oxford: Blackwell.

Hassabis D.K.D., S.D. Vann & E.A. Maguire (2007). "Patients with hippocampal amnesia cannot imagine new experiences." *Proceedings of the National Acaedmy of Sciences* 104: 1726–1731.

Hume D. (1757/1985). "Of the Standard of Taste", reprinted in: *Essays. Moral, Political and Literary*. Indianapolis: Liberty Fund. 226–49.

Klein S.B., J. Loftus et al. (2002). "Memory and temporal experience: the effects of episodic memory loss on an amnesic patient's ability to remember the past and imagine the future" *Social Cognition* 20: 353–379.

Leslie A. (1987). "Pretence and Representation: The Origins of 'Theory of Mind'". *Psychological Review* 94: 412–26.

— (1994). "Pretending and believing: Issues in the theory of ToMM." *Cognition* 50: 211–238.

Lewis D. (1972). "Psychophysical and theoretical identifications." *Australasian Journal of Philosophy* 50(3): 249–258.

Maher B.A. (1992). "Delusions: Contemporary etiological hypotheses." *Psychiatric Annals* 22: 260–268.

Martin M.G.F. (2002). "The transparency of experience." *Mind & Language* 17(4): 376–425.

Mujica-Parodi L.R., D. Malapisina & H.A. Sackheim (2000). "Logical processing, affect and delusioon in schizophrenia." *Harvard Review of Psychiatry* 8: 73–83.

Mulligan K. (1999). "La varietà e l'unità dell'immaginazione", *Rivista di Estetica, Percezione*: 53–6.

— (forthcoming). "Husserls Phantasien", to appear in a volume edited by W. Künne, V. Klostermann.

Nichols S. (2006). "Just the imagination: Why imagining doesn't behave like believing." *Mind & Language* 21(4): 459–474.

— (2008). "Imagination and the I." *Mind & Language* 23(5): 518–535.

Nichols S. & S. Stich (2000). "A Cognitive Theory of Pretence". *Cognition* 74: 115–147.

Noordhof P. (2002). "Imagining objects and imagining experiences." *Mind & Language* 17(4): 426–455.

Schacter D.L., D.R. Addis et al. (2007). "Remembering the past to imagine the future: the prospective brain." *Nature Reviews Neuroscience* 8(9): 657–661.

Schacter D.L., D.R. Addis et al. (2008). "Episodic simulation of future events: concepts, data, and applications." *Annals of the New York Academy of Sciences* 1124(1): 39–60.

Searle J.R. (1983). *Intentionality: An essay in the philosophy of mind.* Cambridge, UK, Cambridge University Press.

Smith A., Li M., Becker S. & Kapur S. (2006). "Dopamine, prediction error, and associative learning: a model-based account." *Network: Computation in Neural Systems* 17: 61–84.

Stevenson L. (2003). "Twelve conceptions of imagination." *The British Journal of Aesthetics* 43(3): 238–259.

Störring G. (1900). *Vorlesungen über Psychpathologie in ihrer Bedeutung für die normale Psychologie*, Leipzig: Engelmann.

Suddendorf T., D.R. Addis et al. (2009). "Mental time travel and the shaping of the human mind." *Philosophical Transactions of the Royal Society B: Biological Sciences* 364(1521): 1317.

Walton K.L. (1993). "Metaphor and Prop Oriented Make-Believe." *European Journal of Philosophy* 1(1): 39–57.

— (2006): "On the (So-Called) Puzzle of Imaginative Resistance". In S. Nichols (ed). *The Architecture of the Imagination*. Oxford: Oxford University Press. 137–49.

Whitfield-Gabrieli S., H.W. Thermenos, et al. (2009). "Hyperactivity and hyperconnectivity of the default network in schizophrenia and in first-degree relatives of persons with schizophrenia." *Proceedings of the National Academy of Sciences* 106(4): 1279.

Williamson T. (2005). "Armchair Philosophy, Metaphysical Modality and Counterfactual Thinking." *Proceedings of the Aristotelian Society* 105(1): 1-23.

Young A.W. (1999). "Delusions." *The Monist* 82(4): 571–589.

Zipf G.K. (1949). *Human Behavior and the Principle of Least Effort.* Cambridge, MA: Addison-Wesley.

www.ingramcontent.com/pod-product-compliance
Lightning Source LLC
Chambersburg PA
CBHW070936230426
43666CB00011B/2461